Beekeeping in Wharfedale

by

Ken Pickles of Addingham
The Wharfedale Beekeeper

BEEKEEPING IN WHARFEDALE

© Ken Pickles

ISBN 978-1-904846-56-7

Published by Northern Bee Books, 2010
Scout Bottom Farm
Mytholmroyd
Hebden Bridge HX7 5JS (UK)

Design and Artwork
D&P Design and Print

Beekeeping in Wharfedale

by

Ken Pickles of Addingham
The Wharfedale Beekeeper

This book is dedicated to Peter Hewitt of Haworth
An outstanding teacher, beekeeper,
honey judge and friend.

INDEX OF CHAPTERS

FOREWORD

As Patron of the Wharfedale Beekeeping Association, I am honoured to write the foreword for this invaluable little book.

I cannot claim to have an in-depth knowledge of practical beekeeping but, like many, am acutely aware of the vital role that bees play on behalf of us all in terms of pollination, not least for food production. The widespread reduction in bee populations over recent years is extremely worrying for us all and we must encourage this little practiced, but extremely important craft.

I can claim, though, to be an aficionado of honey itself and would encourage everyone to buy locally produced honey which will be, quite literally, a taste of what's on your doorstep.

Take time to read Ken Pickles' book ... its contents are relevant to us all, and include useful tips on how even the smallest gestures, for instance planting a few raspberry canes in your garden, can have a positive effect on the bees' environment.

Amanda Devonshire

The Duchess of Devonshire DL

Guardians of the Apiary

*"Chiefly you will marvel at this custom peculiar to the bees that
they neither indulge in conjugal embrace, nor softly dissolve their bodies
in the joys of love, nor bring forth young with their mothers' throes"*
Virgil 70-19 BC

INTRODUCTION

The beekeeping season, like a man's life, is very brief. Therefore, if there is no record of his experiences, his mistakes, his successes, much of the knowledge accumulated over a lifetime will not be available to those who follow him in the pursuit of this wonderfully interesting pastime. It is for this reason that I put pen to paper in the creation of this little book.

The fact that it is documented does not necessarily mean that it is 'Gospel', enshrined in law and beyond error. Perception can be flawed, depending on the glasses being worn! Words are not unlike musical notes on a stave. They are a rough guide; they cannot say it all but they are there to go back to at your will which cannot be done after listening to some eminent lecturer on beekeeping. Having listened to many and done my share it seems to me that many of today's audience of beekeepers don't really want to be given instruction at all but rather they want to be entertained and made to laugh. One can understand that and there is a place for the comedian or one using a joke to reawaken a sleeping class.

A practical demonstration must be one of the finest ways of teaching but the written word must be a close second. This is my objective but I do not mind you laughing if you can find amusement here and there.

As with any pastime there are some purists who are entrenched in their ideas because the 'books' say so. Try to keep an open mind for other writers have been wrong. I remember one saying that grouse did not drink (now there's a thing) and there is no certainty that I have been wholly consistent with accuracy though I try.

These writings cover but a small area of the great county of Yorkshire but they could be applied to much of the United Kingdom, especially in the north. They are based on practical experiences and my interpretation of them. On the balance of probability some of what I say may be right.

The title 'the Wharfedale Beekeeper' was given to me long ago by other beekeepers far and wide that identified me with the valley when serious beekeepers were few hereabouts. I am happy to adopt it during my time.

Recently I was asked if I would like to go back to large-scale beekeeping. Of course I would, I replied and how much better equipped I would be to go about it, and then a twinge in the back comes as a reminder why I now keep so few hives. Each major event in one's life has its season and is replaced by something just as interesting and that, surely, is the wonder of it all.

At the heading of each chapter I have included a quote from the Georgics of Virgil a renowned poet and philosopher who lived over 2000 years ago during the time of Mark Antony and Octavianus. He was born in B.C. 70 at Andes, a small village near Mantua, Italy. He studied bees and agriculture and through the preservation of his writings we are able to gain an insight into his keen observation of the bees. His exquisite and illuminating descriptions are substantially true in most instances and are well worth a read by today's beekeepers.

An Apiary visit by Murial Pitt and school children

'But let clear springs and pools edged with green moss be near and a gentle rivulet swiftly running through the meads;'
Virgil

Chapter 1

Wharfedale

The following is a description of the dale where much of my own serious beekeeping has been done. It is a good example to use being right in the middle of the Pennine hills and sharing so much with those other dales to the north-east such as Nidderdale, Wensleydale and upper Swaledale or westerly, into Airedale; the narrow valleys around Hebden Bridge and north-westerly into Lancashire around Nelson, Colne and the Trough of Bowland where the bees and their keeper have a perpetual struggle with the elements. It is the weather that makes or breaks us as beekeepers and over that we have no control.

The river Wharfe rises above Outershaw in what is now North Yorkshire, flowing down the valley to Buckden, Kettlewell, Grassington, past Bolton Priory then to Addingham and Ilkley. From there it proceeds to Otley and Tadcaster then into the river Ouse before joining the Humber and the Great North Sea. It is a fast-flowing river and equally fast to rise after heavy rain in the upper reaches. The beauty of the dale is unsurpassed in all its seasons and is well documented.

From downstream of Barden Tower it is well-wooded in places. Some of the oaks still standing in the area that used to be the deer park are very ancient. Chris Oldham the Duke of Devonshire's head forester for many years, estimated some were over eight hundred years old, saplings at the time the Normans invaded our land. Many are hollow; ideal for birds and bees to nest in.

This mixed woodland continues more or less all the way down to Otley and though a lot of it is deciduous there are also conifer plantations. In the Barden region there is much heather moor used for grouse shooting and at times access is restricted. This heather extends to the south above the village of Eastby and Embsay and to the east and north of Barden, from Simon's Seat towards Hazelwood to join up with Blubberhouses Moor on the road to Harrogate.

Travelling down the valley on the far side of Kex Ghyll, above Beamsley Beacon, the hilltop heather follows the line of the Wharfe towards Denton and Askwith.

This source of heather honey is very important to the bees kept locally and is of the highest quality, sometimes reaching almost 100% purity of heather. The aroma is powerfully sweet and can be detected a great distance from the hives on a balmy evening after the bees have been gathering for a few days. It is an exquisite honey but not to everyone's taste.

In the early years of establishing out-apiaries I was blessed with knowing the landowners. Without their approval and good-will I could not have built up to the number of hives that I did. Some of these sites I had observed for twenty years before deciding to approach the owner, considering all the advantages and disadvantages for few sites are ideal. The greatest problem for me was access. You may see a perfect site at the edge of a wood, facing south with a barbed wire fence in front of it. The fence itself is a problem but not insurmountable for you can put a stile in. The distance of carrying is the real problem, especially when carrying full boxes of honey.

When young I was warned to take care of my back and like all young people chose to forget the advice. Likely I am wasting my breath telling you the same. I can say this on the matter, when you're lying flat on your back in hospital for weeks on end, you will wish you had not been so hard on yourself.

Where possible get help in the lifting. With two people it is so much easier and the company helps too. I obtained a metal-framed stretcher, bolted at the middle so it could be dismantled. It was so easy to take the roof off a hive and lower this apparatus over the hive to lift and carry with ease. The late Bob Boon of Derby made it for me and it was worth its weight in gold. You could make one out of wood but metal is best. Remember some of those hives on double brood boxes might weigh over a hundred pounds.

The location of my sixteen or so apiaries stretched from Barden to Askwith, many of them being on the Bolton Abbey estate. These were permanent apiaries. Other temporary apiaries were used at heather time.

For many years my heather site was at Barden Scale Farm where the kindly farmer Ronnie Banks and his wife made me most welcome. For their kindness they were amply rewarded with honey. I inherited the site from Percy Ogden of Haworth and he inherited it from another chap who had kept bees there from Victorian times. Originally it was fenced in but by the time I came to use it I had to erect a temporary stock fence each year. Even so, sheep would sometimes get in and push a hive over. I remember seeing one hive completely upside down, still in its lid, with the bees working apparently normally with the floor up where the lid should have been. When returned to its correct position I noticed in the supers how the bees had started to turn the wax cells to hold the honey. They are truly amazing!

Another heather site at Barden was in one of Ronnie's fields below the low reservoir. The advantages of this were it was right in the moor, out of sight of thieves or vandals, and had a reasonable access. Even in those days people unknown to the estate office would place hives just inside the moor gate and leave them during the flow. Then they would mysteriously disappear. I hope the modern-day beekeepers have the manners to ask permission first for it did lead to problems tracing them when the police became involved. You see, the moors are very busy on certain days during the shooting season and beaters don't take kindly to millions of bees coming at them from all directions. The nearer they have to go to the hives in flushing grouse the greater the chance they will be stung.

One year, near Strid Woods Chris Oldham and his foresters felled a plantation. Before they replanted with Christmas trees I watched it for some two years, seeing the ground become covered over again with primroses, cowslips and orchids. I was readily granted permission to place some hives for it was south-facing, reasonably accessible from the car park and well-drained. The area is now planted with trees again and the flowers that adorned that sunny bank are presumably dormant until the present trees are felled.

I already had an apiary in Strid wood itself which acquitted itself very well during the heather time but it did require a bit of brashing out of the undergrowth occasionally. The wasps too were a terrible nuisance at the end of the season. Never have I seen such numbers of wasps attack strong colonies of bees but you see the hives were in the wasps' domain, their natural habitat; the problem was a seasonal event and made any topping up by feeding sugar syrup impossible. The wasps were in such aggressive numbers that they emptied the feeder from within the hives. To even know of this, one had to be there at the right time. Being an out-apiary it would have been so easy for the problem to have passed un-noticed and I think that's true of many apiaries where the beekeeper is an irregular visitor. At least with those apiaries they were safe from vandals and thieves.

For years I considered keeping bees in the walled garden at Bolton Hall. In those days it was completely derelict. Gaining permission from Mr. Ernest Hey the agent and Mr. John Sheard his deputy I started a nice little apiary surrounded by lots of gnarled old fruit trees and scores of sucker plums ideal for walking sticks. It was a warm, sheltered site and the only problem was the drive up for my low vehicle in wet weather.

Another site on the estate was in a bluebell wood near Storiths where Mr. John Croft, the farmer was very helpful. Again, this was not ideal for although I could drive right up to the apiary, I had to lift everything over a dry-stone wall. I never quite got used to that. At times it was very precarious balancing a box of bees on the wall top. Another problem there was badgers, nudging the hives in their attempt to get the honey. They never did but must have caused the bees a deal of anxiety until I placed a fence around them. Bees do not like the vibrations caused by animals such as badgers; it puts them on edge and they will be more likely to sting you or passers-by.

The apiary at Beamsley on Mr. and Mrs Rennards farm was an easy one, perhaps too easy for it was near the lane but in easy flight to the heather as well as the tree blossom in the spring. It did require me to purchase stakes and wire and erect a stock fence to keep the sheep and cattle off but it was worth the effort.

Back across the river at Haw Pike I established another apiary on the farm of Mr. Rowland Carr on what had been the railway line until 1966. The problem here was access, vandals and badgers. Illegal walkers passing along the track would interfere with the hives, giving cause for concern. Percy Ogden, the Snelgrove expert, suggested I put some of our bad-tempered bees along there. Perhaps he was right but I did not do so, thinking it might make the problem worse.

For some thirty years I had a large apiary at Farfield Hall. It had all the qualities required except one. As a static site the colonies did not produce any surplus honey; they had to be migrated to other sites. In the spring they gathered enough to carry them through to the summer and whilst they would fly to the moors they were too distant to gather much more than their winter stores. This site was ruined after a beekeeper placed two diseased hives nearby with American Foul Brood but there had been many happy years spent there as a guest of the landowners and Maurice Riddsdale, the farmer.

The Saw Mill at Addingham was similar. The bees had to be taken to the moors in early August. The proprietor Mr. Billy Brear kindly allowed me to keep bees there for some twenty-five years on a site that his father, Job had kept them. Mr. Brear senior took a great interest in the bees and had no fear of them. He told me how he and his brother would wager on the bees. One would put some in a bag of flour and take them on a bicycle to Beamsley where he would release them. The other would wait with a stop watch to see which hive the white covered bees returned to first. This was a long time ago for Job lived to just short of 102 and what a fine old gentleman he was.

Apiaries at Addingham and Ilkley needed to be out of sight of passers-by where possible. Some people can become very excited at the sight of bees and the bees are vulnerable. One apiary at Ilkley was situated high on a bank on a very valuable building plot. Of course, some people of the more imaginative type assumed I was the wealthy owner. There were houses left and right and behind but there was never any trouble from the neighbours. It is my belief that the beekeeper is largely the cause of trouble with neighbours a bit like dogs worrying livestock where my grandson tells me it's not the dogs that should be shot!

The south-facing side of the river is the most suitable for keeping bees for obvious reasons of more sun. On the north side, above the town, much of the land retains its frost or snow covering in winter, is consequently much colder and being under the brow of the hill, sees little sun. Bees will live there but it is not really ideal. It has its compensations, however, for the woods of Hebers Ghyll stretching towards Netherwood and Addingham, are rich in potential nectar gathering, albeit on the north facing side of the valley.

Dr. John Bates kept his bees at Pawpotts on the road to Beamsley Beacon. They were quite high on the hillside on the edge of the woods and in flying distance of the moor. At one apiary meeting there I pulled off my bee-glove and with it my gold wrist- watch without me knowing the strap had become unfastened. I didn't miss it until later but it was never found despite the use of a metal detector. It had been my father-in-law's and survived Dunkirk then the Burma campaign during the Second World War and that was the most costly Association meeting I ever attended.

One of my favourite apiaries was at the Middleton Lodge Monastery on the south side of Ilkley. The bees were sited at the bottom of a steep bank with a wood behind them in full sun. The brothers were a friendly lot and good company when sitting in their kitchen drinking tea and cake made by Brother Jude. The priest in charge, Fr. Donal Lucey, was very glad to have the bees there for of course they were given honey. This apiary was also in flying distance of Beamsley Moor, a well-kept grouse moor.

One sunny Sunday afternoon I visited the Monastery apiary to find there was a party of school-children under the supervision of a delightful young teacher in summer attire. I ascertained they were from Liverpool and it was clear they were a tough lot. I was asked by the teacher to join them with their picnic and ended up talking about the bees which they all listened to with interest. I was quite taken with the teacher who was so bright and cheery and most attractive. It was not until one of the brothers came across and referred to her as Sister that the penny dropped with a loud clang for she too was a member of a Holy Order. Bees can sometimes lead you into some very interesting situations but it was a grand afternoon.

Some time after that pleasant day I again visited the site to find that someone had rolled or thrown large boulders at the hives damaging several. Perhaps another party of 'deprived' children had visited there. For me it was the signal to close the place down for although it was sacred ground some of its visitors were not ready for the beauty of our valley.

The apiary at Carter's Lane, Ilkley, was observed for some twenty years before I approached the farmer, Mr. Robert Layfield and his wife. They readily granted permission having known me for many years. It is really an extension of Middleton Woods renowned for its bluebells. In proximity to the apiary there are many mature trees capable of giving pollen and nectar such as sycamore, holly, oak and hawthorn. Later in the year there is the useful Himalayan balsam which some purist botanists find invasive. It is invaluable to the bees and can give a surplus crop during the period of the heather flow.

This apiary site was over a fence but south-facing, with a steep climb up a bank. I built a wire stock-fence to keep out the cattle which grazed there and soon had a viable site. At that time, before the trees grew up, there were many interesting wild flowers growing such as I had not seen higher up the valley. I suppose it was the result of the bank not having been cultivated. The proximity of a footpath was never a problem in my tenure.

One thing about this apiary is thirty years ago I would take all strong colonies to the moors at Barden or Blubberhouses in the belief they did better there. With the Carter's Lane apiary I found that the colonies left behind, though not as strong as those taken, produced just as much heather honey without the hassle of moving hives. Thereafter I left every hive on site but I still had to carry heavy supers down the bank to the vehicle.

Mr. Terrence Reilly, a friend from South Armagh wanted to buy some hives so I let him have those at Carter's Lane, continuing the agreement with the farmer. When he gave up with bees the site was offered back to me but I declined for I was reducing my apiaries. Mr. Geoff Halsall and his pal Mr. John Fisher took it over for a few years and had some enjoyable beekeeping there. I believe it has now passed into the hands of strangers.

The Askwith apiary had access problems, being in a wood across a grassy field. It was a case of carrying everything some distance. Its good points were its south-facing aspect and a wood behind to protect it from the north and east. Also the farmers were vigilant so intruders were unlikely. The old beekeepers used to say where there was sheep there was plenty of white clover and that is perfectly true. One of the treats of that apiary was the conversations with His. Hon. Judge Walker for his knowledge of William Shakespeare and the whole of Henry V, including the bees, brought pleasant entertainment.

At Weston, Col. Dawson, allowed me the use of a small piece of land near the church but sadly this was prone to evildoers who vandalised the hives.

Most of these sites in the valley bottom allowed the bees to work the low ground and the high ground which was a great saving in labour at heather time.

From Ilkley down-stream through Burley, Otley and Arthington the conditions for keeping bees improved for there was a greater chance of a crop of surplus honey for the beekeeper. More farm crops such as oilseed rape and field-beans helped in one respect and there were more flowering hedges. Dry stone walls secrete very little nectar!

Up-stream from Ilkley, through Addingham, Bolton Abbey, Barden, Burnsall and Grassington the likelihood of a crop of surplus honey diminishes except at heather time. Just beyond Barden the Craven Fault causes a change in rock types from Millstone Grit to Limestone, from a more acidic soil to one more alkaline. Beyond Grassington up to Arncliffe it is still limestone but here and there, on the hill-tops there are outcrops of millstone grit and more acidic soils which give rise to heather again such as at Boss Moor.

Deciduous trees are not as numerous as in the lower reaches of the valley and during poor seasons bees would struggle to survive without assistance from the beekeeper, unless, of course he left them all their honey, which any decent beekeeper would. Grass woods provide a good source of nectar and pollen and in the spring there is a good supply from hazel. Later in the season the yellow ragwort redeems itself somewhat by giving nectar when there is little else except late white clover, a fickle plant at the best of times for it quickly dries out.

Not all parts of the valley will sustain bees and produce a surplus but it is possible to keep bees in those places or at 600 or 700 feet provided no honey is ever taken but it is not recommended for they live on the edge of survival most of the time.

Bees require lots of water in the breeding season and the valley has a plentiful supply even in dry seasons. Bees prefer stagnant water and will have their regular watering holes which may not always seem hygienic to us. Rain water lying atop of cow pats draws them and sphagnum bogs where they will draw it up through the moss. Sometimes the early morning dew satisfies their quest or puddles in the track. In very hot weather much of this water is used to keep the hive cool distributed in empty cells about the combs by those bees designated for temperature control.

In the length of the valley pollen is abundant and I have never known a shortage. On the contrary I have known the brood combs to contain more pollen than the bees need thus reducing open cells for the queen to lay in. No doubt the bees know why they are doing this so I tend to leave well alone. If you remove combs of pollen they will go mouldy in a few days without the attention of the bees and that seems a shame for a lot of work goes into its gathering.

The disaster at Chernobyl in Russia was also a disaster for Wharfedale though we didn't know it at the time. Radioactive fallout was quite heavy as it was in other parts of the United Kingdom. The contaminated clouds were swept this way and the heavy rain brought it down. The readings varied considerably but were greater on the hill tops than in the valley bottom for some reason. Seemingly moorland plants such as heather, crowberry and bilberry can take up the caesium through the roots. This is gathered from the flowers and ends up in the honey. The sheep and grouse that ate contaminated heather also became contaminated. Prior to the nuclear accident there was no caesium problem on our heather moors for it wasn't there. It will take a long time to go away. The experts say there is no risk to human health.

I have become allergic to heather honey since the Chernobyl accident. Within a day or so of eating any I will come out in a sore rash down each side of the nose. Perhaps this has occurred naturally because I have eaten so much in my life but I do wonder.

Over a fifty-year period I would say of Wharfedale that for most of those years the bee seasons have been mediocre but at one end of the spectrum there were a few outstanding years and at the other extreme there were some devastating years. When you think of it, this is much like the life of the average human being so perhaps we have more in common with the honey-bees than we realize. As I have often said, if beekeeping were too easy, everyone would be keeping them.

Wharfedale, because of its great beauty and its geographical position in relation to the cities of Leeds and Bradford has, over the past forty years attracted large numbers of people from those cities and further afield. Some of the locals refer to such people as starlings; i.e. flying off in the morning and returning to roost in the evening. Of the vast number it seems very few understand or even want to understand the countryside which is an awful shame and perhaps even detrimental to the valley in which they have set up their home. Land taken for house-building need not be totally lost to bees and birds if suitable trees are planted. I mean, even the smallest garden could have a Victoria plum and apples and pears can now be grown on dwarf root-stocks. Reputable garden centres stock all these plus many more plants useful to bees.

It is not really sufficient to become just a beekeeper even though you cannot put a price on the value of your bees to your local environment. Once you take up the craft there is much more can be done to improve things for the future and trees and plants are the main things that come to mind. Urban people too, in time, might revert to being country people though I think it unlikely unless it becomes a necessity for survival but one never knows.

Wild garlic in my Apiary

Peter Hewitt watching bees work oilseed rape

'In wandering among the flinty rocks, have they torn their wings, and voluntarily yielded up their lives under their burden: so mighty is their love for flowers, and such their glory in making honey.'
Virgil.

Chapter 2

An Outline of the Craft

Some beekeepers like some politicians or theologians can become so fixed in their ideas that they will not consider change or the fact that they might be wrong at any price. They have accepted a certain doctrine and because it has been practiced that way since Henry was King they will not reconsider their ways. I suppose we've all been a bit like that at times. I mean, not so long ago, no one would have batted an eyelid about cruelty to chickens in battery cages or a bit of badger baiting or otter hunting for sport.

Yesterday morning I watched with pleasure a vixen as she lay underneath my disused hen-cote watching me. She rose slowly, stretched (still watching me), turned her back and trotted off as only a vixen fox can. She was beautifully coloured, her black tipped ears, her white tipped tail and her red-brown coat in perfect condition. I felt privileged to have seen her. No doubt she watched me every morning as I visited to feed my pea-fowl.

Now, you're thinking what has this to do with bees. Quite a lot actually! You see, once upon a time I would have been reaching for my gun to shoot that fox or even worse, setting down three cruel gin traps (rightly banned in 1958) with a suitable bait to catch her.

I speak as a man who has changed his mind about certain long-held views, not just in the execution of a fox for trying to survive but in the way I now keep bees. This chapter was originally to have been placed at the end of the book but I changed my mind about that too. You see, my reconsidered views on bee management are now so very different from the orthodox teaching that I think it important that any newcomer to the craft should hear what they are before digesting any more of the trendy claptrap that has been propounded in man's ever expanding greedy desire to take more than his share of nature's bounty. Self extermination of the race is surely inevitable unless people do change their ways and, whilst some may not like it, it does include the beekeepers. Because practically all honey-bees in Britain are now under their care and suddenly the Nation's spotlight is on them and their management, a little silent thought would not be amiss on the welfare of the bees and how to improve their lot.

Rough handling of bees is seen all too often at apiary meetings. It is not acceptable. Bees are livestock like any other and should be treated gently for they are gentle creatures by nature unlike the people keeping them who often seem in a hurry to get the job done at the expense of the bee's welfare. Take your time and enjoy it. I don't mean expose the brood nest overlong or even keep the hive open longer than necessary for your purpose but do use your eyes, take note and learn. Remember, the fun is in the chase and beekeeping can be like a love-affair and often is.

Still on the subject of cruelty and needless slaughter of honey-bees I would like you to imagine you are in a standard apiary with a standard instructor who will demonstrate how to inspect a hive. We will comment on the reasons why at the end.

Firstly, he will puff a little smoke in the entrance then remove the hive roof. With the hive tool he will free the crown board on one side puffing in a little more smoke. Because the board is well propolised he has to work all round it with the tool. Because he allowed the partly raised board to return to the top of the super he has killed some half a dozen bees.

The hive comprises a single brood box and three supers. It is the second week in June, quite warm with no wind and there is honey in the supers. Keen to have a glance at some freshly gathered nectar he frees one of the flank frames then with the hive tool frees several more. This operation kills another half a dozen bees.

Using the hive tool he frees the top super and makes to lift it but a frame is stuck so the box returns to its original position killing a dozen or so bees. He tries again and places the top super in a diagonal position on the upturned roof. So far so good!

The next super has lots of brace comb on top of the frames, much of it filled with nectar. Our expert beekeeper of some twenty years uses his hive-tool to remove it, first smoking the bees again. This scraping along the frame bars kills or maims another half a dozen bees. This second box and the third box are similarly removed killing more bees in each manoeuvre.

Coming to the queen excluder, the hive-tool comes into play again. It is removed, shaken and placed to the side. More smoke is given.

The brood nest looks good from the top but our instructor will have a look anyway.

Loosening a flank comb with the hive-tool he tries to lift the comb out. Having been on a single brood box during the sycamore and hawthorn flow, the bees have drawn the comb to its maximum and filled it with sealed honey.

Because the comb is irregularly shaped the bee-keeper takes it by the lugs and jerkily forces it up the hive wall crushing many bees on the next comb and the wooden wall. He places this at the side of the hive, loosens several more combs and takes one out with sealed brood.

This is examined for disease, for eggs which tell him the queen is laying and finding eggs he sensibly returns the comb exactly where he took it from. Shaking all the bees off the flank comb into the hive, he replaces it where it had been before. He re-assembles the whole hive exactly as it had been. By now he has back ache, is very hot in his suit, veil, gloves and wellington boots but he feels satisfied. Why?

It was a fine sunny day with a nectar flow in progress. He soon put the bees off their work by his actions. He killed bees un-necessarily. In fact the whole operation was un-necessary.

The bees, not liking being handled in this way, have now started to encourage the queen to lay in the queen cell cups. In a week's time or thereabouts, the first of those cells will be sealed and that colony with throw a swarm if the weather conditions are right.

So what did the beekeeping instructor gain by going into the hive when he did? for the bees, absolutely nothing; for instructing the novice beekeepers little that couldn't have been done at another time of the year. Remember, his actions have provoked the bees to prepare to swarm. If he knows about it, he is forced into a lot more work. If he misses the signs and the swarm issued, there is still a lot more work and he may lose the swarm. Had he left the colony alone the chances are the bees would have completed the honey flow without thinking of swarming and also the crop of honey would have been greater. Once bees are preoccupied with the swarming impulse their capacity to work is reduced and less honey is gathered and once the instinct to swarm has been activated it is difficult to stop. Attempts to do so often end in failure and can wreak the colony. Sometimes, however, if bad weather comes and remains a while, the bees will tear down their queen cells and forget the whole idea. That is very pleasing when they do, saving much work.

That described above is orthodox beekeeping and has been so since about the 1870's when wooden hives with bar-frames started to come into regular use. This coincided with another phenomenon which greatly affected the lives of country people and still does so to digress for a moment.

Just as the change from skep hives to bar-frame hives was taking place the breach-loading gun was invented. This allowed game to be shot at a much faster rate and led to the big shooting estates and the development of grouse moors for the same purpose by the wealthy land owners. Those same people were also in Government and soon brought out draconian laws (still on the statute book) to protect their interests.

The Poaching Acts of 1829, 1830 and 1831in effect criminalised all country people who had relied on the odd rabbit or hare to supplement their diet, for someone had to deter the 'peasants' from eating the pheasants or the grouse, a bird which the 'peasants' had always considered inedible, tasting of moor and bog. Ralph Waldo Emerson, the American writer, said the English Game Laws were the most oppressive laws ever perpetrated against a free people. The by-product of this was they helped populate the colonies, especially Australia.

Somewhere there is a balance in the exclusion of people (all people) from wilderness areas for periods of time, especially during the nesting season whether for shooting purposes or not for it does give wild-life a better chance without people and dogs wandering everywhere.

Had it not been for the breach-loading gun and the red grouse the sheep would likely have cleared the heather off long ago as they have in many un-keepered, un-protected moorland areas of England and Scotland! So you see, another by-product of grouse shooting is the massive expanse of heather moor of all ages, just right for the bees, for every loss has its gain! Gamekeepers were employed to burn the heather, something I have often taken part in myself. As one keeper friend, Walter Flesher of Burley Moor told me, 'no heather should be more than ten years old for grouse to feed on' but of course some leggy stuff needed to be left for nesting cover.

Mr. Stitt senior, the head-keeper at Bolton Abbey for many years did not believe in burning heather, seeing it as well able to regenerate itself. The result is there are still some fine patches at Barden as tall as a man. His son Bill Stitt who also became head-gamekeeper did believe in burning which is why the moor is so good for grouse and bees to this day, the tradition being carried on by Mr. Harold Allison and subsequent keepers.

This grouse-moor management has worked to the advantage of those beekeepers in reasonable distance of a well keepered moor and provided it is not overgrazed by sheep there should be no reason why these moors should not provide food for bees for ever. Sadly, heather moors on Addingham Moorside and over the hill at Riddlesden were ploughed up after the war, assisted by Government grants, destroying large areas of natural habitat for birds and bees.

It was the attraction of the heather honey and necessity after illness that caused me to increase the number of my hives for very few people were taking advantage of this rich harvest. In Burnsall there was one beekeeper, Mr. Lodge but no other until myself at Addingham. At Ilkley there was Frank Beatham,

Laurie Illingworth, Geoff Halsall, and Frank West. At Burley there was Dr. Green and at Otley Norman Gravell and Mr. Mallinson.

Knowing all the farmers, landowners and gamekeepers I had no difficulty in acquiring out-apiaries for heather honey production. I knew the risks involved for I had been at it already for a number of years but I was fit, strong and above-all had the energy and drive to see it through. I did not have capital so I made my own equipment and that in itself was satisfying and very therapeutic.

Taking one or a hundred hives to the heather moors is a gamble. No one can be certain what the season will bring. Perhaps the risk factor is the attraction for some. I don't know for I was never a gambler and would have much preferred a guaranteed crop. Alas, nature is not like that; she likes to keep us guessing, sometimes right until the last minute, when, after atrocious weather, we see September come in, see the heather is fading and know there is still nothing in the hives. All is not lost, however, for the temperature rises and there is a fortnight of gloriously warm, sunny weather. The bees go into a frenzy to turn the tables on their precarious existence and the supers fill up and even the brood nest is packed with the liquid gold wherever they can find empty cells in which to place it. On extremely heavy flows I have known bees climb up the hive walls and draw wild comb under the roof when their supers were full up.

September flows on that scale are rare. Usually the flow comes at any time from the middle of August onward. It can be very humid about the time of the glorious twelve, the opening day of the grouse shooting and sometimes thundery but humid weather nearly always brings in the nectar for it does not allow the flowers to dry out.

In the cleared forest areas there are often large patches of rose-bay willow herb, commonly known as fire weed on account of it growing where the foresters have had a burn. This blooms at the same time as the heather and the bees will work that. One can tell by looking at the flowers or watching the entrances to the hives where the blue pollen from the herb is being taken in, the heather pollen being a dirty grey colour.

If the hives are taken to the moors before the ling is properly out, in July for instance, the bees may be lucky and find enough Bell heather to supply their needs but there is no quantity of that here like there is in Scotland or the North Yorkshire Moors. Also at that time, where there are sheep, there are masses of wild white clover. It is said this secretes better on chalky soil but even here, provided it is not too hot and dry it does produce some food for the bees.

Another source of food which can occur in July is honey-dew from the pine tree plantations surrounding the moors. This stuff is a secretion from aphids, quite dark in colour but palatable and enjoyed by many. In the jar it looks like old engine oil but if held to the light has a slight purplish hue. It is quite acceptable in the 'dark honey' class and I have won prizes with it long ago. I am told the Germans greatly value it, there being much of it in that country.

Another thing that is very noticeable at the heather time at Barden and Bolton Abbey is the rich orange-red colour of the masses of propolis gathered off the plentiful pines. Those colonies that draw a curtain of the stuff across the entrance and below the first frame (warm way) I suspect are descendents of Caucasian bees I acquired years ago for they had that marked characteristic. Propolis so gathered is handy to cut off and place in a jar for 'medicinal' purposes; i.e. tooth ache or sore throats.

In the case of a troublesome tooth just place a lump against the tooth and press it into place. Remove before you go to sleep in case it falls down your gullet and chokes you. Do the same for a sore throat. It is very effective.

For the month of August and for some ten days into September Wharfedale's heather moors would provide sufficient fodder for many bee hives. Certainly hundreds; perhaps thousands but the period would be short, the apiaries would have to be spaced well apart to minimise the risk of disease and the apiaries themselves should not be overpopulated and not in lines, all facing the same way. Perhaps a dozen hives is the optimum number for minimum risk, facing different directions in the shelter of a wall, with a stock fence round to keep cattle and sheep off. They are better slightly off the ground with a flat board at the entrance to stop the grass growing up and blocking it. If the roofs are flimsy and light it is wise to place a heavy stone on each. For myself I prefer a deep, heavy roof. It saves messing about with stones.

My good friend Eric Dixon of Lothersdale kept his bees in his garden on top of a hill. It was a cold, blustery place but the bees survived and usually brought him home a crop of heather honey from the nearby grouse moor. He had no cause to move his bees to any crop but was satisfied with the little they gave. I suppose that's about the best he could expect.

'Therefore, though a narrow term is their lot yet the race remains immortal, and through many years the fortune of the family subsists, and grandsire of grandsires are numbered.'
Virgil

Chapter 3

Why Become a Beekeeper?

For some people beekeeping can become an over-riding passion in a similar way that some men or women are drawn into political, religious, philanthropic or just plain old-fashioned bird-watching organisations. For me and others I have known bees became a sort of fever. I couldn't get enough of them. It came about gradually from an early age until my leisure hours were mostly spent with them. I ate, drank and thought of beekeeping whenever I could. The house smelt of bees, the car smelt of bees and according to my close friends, even I smelt of bees. Other farmer friends smelt of sheep, cows or pigs. So did their vehicles and houses. I knew which I preferred. Beeswax and honey is a heavenly smell, like the bees that produce it and no beekeeper should feel guilty about that. Like my farmer friends it couldn't be avoided anyway.

The working with, and the desire to understand honey bees, has always been the lure for me. Honey production, whilst much prized for the quality food was a secondary part of the interest for its handling was sticky, messy and very hard work due to the constant lifting of heavy weights in the form of boxes of combs or buckets of honey. I would far rather be out at one of my apiaries in the clean air tinged by a little wood-smoke from the smoker, learning about the bees by quietly watching their activities, collecting a swarm or just admiring the other aspects of nature in all her glory.

I do not say honey was not useful, valuable and always welcome as a gift to friends but the disposal of it by selling did not excite me. In the case of selling to shops it seemed many shop-keepers did not realize its true value, thought they should get it cheaper than it was worth and were not always loyal customers if some other unscrupulous beekeeper came along and undercut your price. Business never interested me like it does some and though I sold some honey I never classed myself as such, tending to give away as much as I sold.

If the true value of British honey was on the label I think only the very wealthy would be prepared or able to pay for it so do use it sensibly and don't cast your pearls before swine. Remember a great many lives were lost in its production.

From the outset of them taking up beekeeping as a hobby I have known some of these enthusiasts unable to wait to steal the honey from their few hives and get it into labelled jars with their name on and see it handsomely displayed in the local shop. Of course they cannot maintain the supply and it has gone in no time.

Others I know of, who started out as beekeepers, saw the sale of honey as a lucrative business and being unable to produce enough from their own bees, chose to buy honey in bulk from other amateur beekeepers and sell it under their own label; others, being even more ambitious as business people abandoned the beekeeping altogether and imported foreign honey, selling it as such. As Jean Brody said, 'For those who like that sort of thing, that is the sort of thing they like' and so it is.

For those whose backs remain strong I can think of no greater occupation than working a bee farm. Like all men and women working on the land for a living they are very brave for the risks are tremendous and not just to the back. Having more hives to look after increases the risk of injury, of disease among the bees and of course the dangers to the bank balance if a succession of bad seasons hit the enterprise. Other than that, for the single-handed man running a few hundred colonies, he works mostly alone, at peace with the world and his bees that he truly cares and worries about, ever watchful to their needs and the dangers that beset them and with them shares a freedom that few other humans would really understand. In effect, he becomes not unlike a bee in his thinking though not in appearance. However, I have known some dog owners start to look like their dogs.

I think the 'fever' description is most apt for the real professional beekeepers I have known and it sets them miles apart from the amateurs in their attitude, their knowledge and understanding of what is at stake. They are the few, mostly in the Bee Farmers Association, who really do take a gamble each season. For the amateur, if his one or two colonies die out, it is not critical to his welfare and if he is wealthy he can soon purchase some more. For the serious beekeeper, to lose many stocks can set him back several years with his business and may even close him down.

Some amateurs have viewed with suspicion or envy those who try to eke out their pension or make a business of full-time beekeeping. Human nature being what it is that is sad but understandable. Certainly, some professionals nearly die knowing what some of the amateurs get up to in imperilling the health of their bees. There is very little that can be done about some of those practices other than to try and educate the novices from day one that keeping a stock of bees carries responsibilities too.

One old time beekeeper I know tells me where for all his beekeeping life he was the only one he is now surrounded by nine other apiaries. This sort of thing is a real worry but there is nothing to stop such proliferation. It is a case of putting on your tin hat and waiting for the fad to fade as well it may for there is not unlimited food for bees in every area of the valley. If there are too many stocks in a given place none of them will be able to gather sufficient to survive on. Time will tell on that score. It is the same for all livestock though people tend to forget this, especially those in government who have allowed unlimited access to these already overcrowded islands. Time will also tell in that contentious case whether the experiment was wise but as a beekeeper, if you choose to flood an area with large numbers of colonies, I think you will quickly learn what I am talking about.

Ideally apiaries should be two or three miles apart so that no single one is monopolising the food supply; from a disease point of view, the further apart the better. Even in apiaries for over-wintering where stocks of thirty or forty are placed close together for the convenience of the beekeeper, widespread fighting and robbing can occur throughout October into November when the first frosts reduce all to calm.

We must always assume that some disease is present in most colonies all the time, particularly Nosema and Acarine. Apiaries located in damp places tend to produce Acarine disease in poor seasons.

Colonies that are living largely on heather honey, if confined for weeks or months in a severe winter, may suffer dysentery and void themselves in the hive. This is one of the characteristics of heather honey and probably explains why some beekeepers like to top up with sugar syrup at the end of the season to obviate this happening. The contaminated combs will be cleaned up by the bees encouraging the spread of Nosema.

This winter of 09/10 the bees were confined for over six weeks and the front of the hives were plastered on the first day they could fly for a very short time in early February and that was for only a half an hour. They have been confined ever since and now it is the 28th of February. I find it amazing that some beekeepers who have messed about with bees for over twenty years do not understand why this spotting of hive fronts comes about. Either they have not glanced at the hives in winter or have not read a book on the subject. The 'spotting' may just be simply the result of the long confinement and if it is connected with disease there is nothing can be done about it at that time of the year. The bees will deal with the situation as the weather improves. Always, some will live and some will die. They are all different, some having resistance to most things, others weak, unsuitable and prone to collapse. Didn't Charles

Darwin mention such things?

Bees in transit will often dirty their combs, presumably through fright, which is one good reason to avoid moving them at all. Static sites in a good area must surely be more desirable? I can say that for I have done an awful lot of migrating of bees and there is always a price to pay for it. With any form of disease, prevention is the best thing but of course many young, ambitious beekeepers, ignorant of the cost to their bees, will drag their undersized charges all over the country looking for the rainbow and that pot of gold when season for season, they would have done better leaving them at home.

When you migrate into other peoples areas you have no idea what problems your bees might encounter. I have found derelict apiaries where bees have died out from heaven knows what. These attract your swarms should you lose them. Most certainly your bees investigate them and should there be disease it is so easily transmitted back to your healthy hives.

Migration to the heather moors whilst very attractive carries the same risks as described above but in the times in which we are entering, with the increased number of novice beekeepers keen to have a go, the threat to the health of the bees is greatly increased. Sadly, many of these people will only learn from their mistakes and will not be easily dissuaded from their actions and for the time being at least we will have to put up with it. Nature usually takes a hand with these problems anyway so it's no use worrying.

The teaching of anything can bring great pleasure to the teacher and beekeeping is no exception. To have a class of students of all ages enthusiastic to learn about the craft rekindles something of that zest the teacher had when first made aware of their wonderful life style, in my case by my father. I exhort all beginners to gather all the knowledge they can for there is a lot of it, then to sieve it until you retain that which is essential to your own particular needs. In other words, don't get bogged down in the trivia. Attend all the classes, talks, demonstrations you can then use your own intelligence to sort out the wheat from the chaff for sadly, there is still a lot of chaff.

Years ago I taught many people bees without charge, giving my time and energy to ensure their success. With some I even supplied them with bees, at no cost. On the whole I think that was a mistake. From my experience people who are given bees don't always value them; the easy come easy go attitude I suppose; one of the failings of this generation. If you have to dispose of bees sell them at a fair price. You don't have to retain the money for yourself; give it to the Sue Ryder or Marie Currie Foundations. After all they even care for ailing beekeepers.

Many of those I taught kept bees for many years but give it up for one reason or another; failing health on their part, too many disappointments with the bees or a bad stinging experience or because it was not lucrative enough for the effort put in. Neighbour problems are another cause of people giving up for they have not the time or patience to deal with an out apiary which is the only alternative when that occurs. Some of them, alas, want it on a plate; everything provided, expecting sites to be offered to them. Good sites are hard to come by and should be treasured.

Just occasionally there will be one student who really does get 'bee fever' and there is no holding him back. I say him because, though I have taught many ladies, I have never had a single one who became fanatically engaged. It was the same with teaching the bagpipes mind. They are physical too so maybe that is the reason. Lifting boxes or complete hives is hard for a man so it must be so for a lady but is that politically correct to say that anymore? I know in nature there is always the exception to the rule. I just wish I had met one of those strong ladies to give me a hand.

One of my best, most recent students (some ten years) is Anthony Kirby, currently training for the priesthood. Initially we shared the interest in the highland bagpipe but then he became a beekeeper and now runs a sizeable apiary some thirty miles from Wharfedale. He soaked up knowledge like sphagnum moss soaks up moisture and put it to good use.

His apiary is the tidiest I have seen; as tidy as the late Douglas Morris of Bramham and he was pretty tidy. Careful and methodical in all his dealings with the bees Anthony gets honey when no one does. This last year, 2009, he went to buy 500 jars from his local supplier. He was asked what he wanted the jars for and when he said honey the supplier was amazed saying he was his only customer. He was even more amazed when Anthony told him the amount. You see, even in bad years, some areas can surprise us. The honey was a nice blend of balsam, willow herb, white clover, beans and some rape, which seems to be grown all the year round in that area. It does pose problems and it did for Anthony but he overcame them but it means the loss of some combs.

Anthony's success is due to a rapid spring build up of the bees in a more temperate area of Yorkshire blessed with plenty of early forage.

Generally, in Wharfedale, there is a very steady spring build up due to the wet and the cold winds of March and April. This means, unlike Anthony's stocks, the bees are not strong enough to take full advantage of what becomes available when May comes along. However, I know of one or two places in the valley where bees tend to build up much better but they are downstream on warm sheltered sites.

After attending the elementary classes provided by Beekeeping Associations at fees varying from £20 to £50 a student should have a grasp of the basics. This should be followed up by reading every book on the subject and getting out with the bees, preferably other peoples. Leave your own alone and keep them stress-free. As regards the books, remember, the bulk of them are written by south-country beekeepers living in a climate very different to what prevails in these unpredictable, wet, Pennine hills.

There is no substitute for practical experience and I make no excuse for repeating this. Like everything else in life you must be prepared to make mistakes. Don't worry too much about that either. Some of the more experienced beekeepers will be glad to assist when you hit problems. I know my phone used to be red hot from such callers in the summer right up to bed time. I confess I am glad I rarely get troubled now but no doubt one or two I taught carry on the good work and pass their knowledge on.

Classes are fine but for me the one-to-one approach is far superior as a method of teaching whatever the subject might be. Of course, that is not practical during this present upsurge of interest caused by the media but, like everything else, it will eventually resume normality.

Assistant Harry Jeffrey of Silsden

'First a seat and station must be sought for the bees, where neither winds may have access, for the winds hinder them from carrying home their food.'
Virgil

Chapter 4

Nectar Sources and early management

From Otley upstream it is rare to see any fields of yellow brassica crops, whether oilseed rape or mustard but just occasionally a game strip will be planted containing these plants. Other than that, we can be sure to get a quality honey, uncontaminated by brassica even if there is not a lot of it.

Willow is perhaps the first to yield in any quantity in late March and April but the colonies are not strong enough to take full advantage of it. It coincides with the tail-end of the snowdrops and crocuses in gardens and in the woods, the hazel, and alder catkins. The bees get much pollen from these and some nectar. The cheery yellow coltsfoot does not go un-noticed by the bees and there is more of this about than people suppose.

With the dandelions coming into bloom it is usually a good time to place supers on the hive. I say supers because you might as well get several on in one go For one thing, giving room early inhibits the swarming impulse and allows lots of storage room for nectar should it be a bumper year. If the supers are in your shed they are no good to the bees.

The honey from dandelion is very good but has an unpleasant smell. Rarely will you taste it for it is required for the build-up of the bee colony. Bright orange-yellow pollen going into the hive will tell you they are at it. The dandelion plant is one of the most valuable to man as well as the bees but few choose to regard that anymore. For back pains, urinary problems and just to be eaten in a salad it is a real benefit which is why it is so common, like the wild garlic or Ransoms which grow so prolific in our woods, now more or less neglected but a great purifier of the blood, I am told. The bees work them too. In addition, as I have already said, they will cause the Varroa mites to die without harming the bees.

The sycamore blooms over a long period because there are many varieties of this useful Acer tree. For those of us with bees in Wharfedale and district it is perhaps the most important source of early nectar. If the bees are of a good strength during its blooming they can get enough honey to allow the beekeeper to take a few combs. It is followed by the May blossom. This needs calm warm weather for the bees to have a real chance. For me, it is one of the nicest of the spring honeys but the weather is often cold and wild, preventing the bees making use of it. As with all these flowers mentioned, being in the valley bottom means that the flowers there die off first but higher up the hillside there is a succession of bloom which may last into June with the May blossom. Alas, when that dies, we know there is no major source of food for some time. This has been called the 'June Gap'.

It is at this time some unwise beekeepers, delighted to see one or two boxes of honey on their hives, remove it and resort to the heinous crime of feeding sugar syrup. In our ignorance and stupidity we have all done it but it cannot be right to do so. The honey the bees have gathered is required to make strong healthy bees. Sugar syrup can never do that. It is a refined substance, treated with chemicals to achieve that state. Honey is the bees' natural food containing much that we do not yet understand but we know they are far better for it. In fact, some beekeepers believe the breakdown in the bees' health is in part due to this widespread feeding of sugar. I hear it's not very good for human beings either.

The one occasion when it is justified to feed sugar is if the bees are starving as in a bad season, for dead bees are no good at all but I do plead for the bees that novices, and experienced beekeepers will leave more of the honey on the hives and so put this ridiculous, greedy practice into the history books. More colonies die in April than any other month which is a tragedy for they have somehow struggled through the worst months of winter to see the start of spring and the life-giving flowers only to fall victim to starvation because the beekeeper took more than his rightful share in the autumn.

From about the tenth of May onwards some colonies may swarm. This is perfectly natural and nothing to get too excited about though I do confess to finding swarming very exciting. It often occurs during a honey flow when it gives the swarm a better chance of survival. Now, this may not be very convenient for the beekeeper, his only object being a large crop of honey for he knows that the division of the colony may reduce that crop. This may not necessarily be so for the swarm, if caught, hived and given the supers from the parent colony, may fill them if the flow continues.

The stock from which the swarm came should produce a young queen and if the swarm has been hived alongside it can be united to the young stock at the end of July just prior to the main flow from the heather. This, of course, entails more work in finding the old queen and despatching her and joining up with a newspaper. The Daily Telegraph is about the only paper left of the right size for this purpose using one sheet. I always found the bees liked the Guardian best! When uniting by this method I do tend to pierce the sheet of paper with the hive tool to allow the common scent to develop more quickly. Within twenty-four hours much of the paper has been thrown out of the hive entrance and can be seen on the ground. When the bees are related there is less chance of fighting. Indeed, when I used to unite related skeps of driven bees there was no fighting at all and that was without any aid such as newspaper.

During June the bees sort of 'tick over' which some beginners may find hard to understand. Often the best weather comes at that time but the main sources of nectar as previously described are finished. The honey left on the hive tends to reduce with the increased brood nest or with swarming for a swarm fills up with much honey before setting off on its adventures. This is required to draw out the new combs and sustain them should the weather take a turn for the worse.

All through the swarming season, if you have been at the craft some years, it pays to set up one or two 'bait' hives in the apiary in an elevated position. Should a swarm issue unexpectedly it may enter one of these 'bait' hives.

The hive itself should contain a full compliment of clean, used brood combs (not combs that bees have died out on. Disease-bearing combs merely perpetuate the scourge and that will cause you no end of trouble later on. Diseased combs, spotted with faeces containing Nosema spores are better burned (hang the expense ñ cheap is dear, you know).

A normal crown board with a heavy roof should be placed on top and the entrance block should be set to a very small opening. Bees prefer that to a full width opening. When swarming they seem to have that secrecy air about them borne out over hundreds of years of being messed about by human beings.

So often you will be in the apiary when a swarm issues. Its first indication is usually the noise if you're wearing a veil. You will then look up and watch in fascination as this huge cloud of happy bees swirls about. First you think its going on the pine tree, then it switches to the willow, then it has a look at the hawthorn. By then you've gone looking for a skep to put it in but when you return the excited hum has gone and with it the swarm. They've fooled you well and truly and absconded you know not where. If in the open countryside you will be lucky to find it so save your precious time and get on with something else. Your loss may be someone else's gain.

The loss of a swarm can be a blessing in a bad season for it is fewer mouths to feed so don't spill too many tears. This acquisitive society has to relearn that it has to let go of some things with good grace and a swarm is one of them. As an aside, I still believe it is the law that a swarm issuing from an apiary, if followed and kept in sight by the owner, still remains his property. If he should lose sight of it just for one moment he loses his property rights in it and ownership falls to the person who catches it or on whose land it is.

By the end of June most swarming is over and done with and the apiary is looking a little bit untidy with different sized hives dotted about all over the place, some containing swarms, the others the parent colony. However, there may be another burst of swarming activity in July. All this can be put right at the end of the season but until then colonies are better left as integral units. United stocks tend to do less well than those left alone.

In July things start picking up a little nectar-wise. The Rose bay willow herb, that strikingly tall plant with the purple-red flowers is so useful to bees with its thin, watery-white nectar and its blue pollen. Douglas Morris the bee farmer of Bramham averred it secreted better in the industrial areas where it was prolific on waste ground. He believed it was better there than in the countryside because of industrial fallout from the factories and mills enriching the soil. Perhaps he was right. On the old railway line between Ilkley and Bolton Abbey there was tons of it where recurrent fires encouraged it to thrive.

Other July flowers incredibly useful to bees are the blackberry, second cut red clover after sillaging and where there are sheep, the wild white clover. By the middle of the month, if you're lucky to have them, the lime-trees are in bloom. These magnificent trees can supply a number of colonies with food which is very good quality. If the weather is very hot they tend to incite aphids to secrete honey dew thus attracting honey bees and many Bumble-bees. The latter can be seen on the ground in great numbers, apparently inebriated. In the evening hedgehogs are drawn to the feast of sweet Bumblebees.

If in reach of the moors your bees may find an additional moorland source from the Bell Heather. I have only had advantage of this in Scotland but it is a fine honey though not to be confused with that from the Ling or CallunaVulgaris. By the time the Ling is fully out by about the 12th of August, the bell-heather is mostly over.

Himalayan balsam is a truly beautiful and interesting plant which came to us via Australia, the seeds travelling in the bales of wool from that continent. When the wool was washed out the seeds entered the British water courses and first germinated in the industrial heartland of the heavy woollen district. From there it has spread all over the country. I have even seen it in Caithness and Ireland.

It comes into bloom in July and lasts until the first frosts in November. Bees can work it for pollen and nectar in the rain because of the position of the large, bell-like, pink flower which hangs at such an angle that the nectar is not washed out as in many other flowers. It is also a favourite of the Bumblebee.

The pollen is whitish and the nectar opaque and those colonies in the valley bottom will work it on those days when flying to the moors is prohibited by weather. They can get a good crop from it though it is stiff honey and not easy to extract.

During July make a mental note of your strongest colonies fit to take to the moors. For convenience a single brood box with two or three supers is ideal. Additional supers may be added later but if that is the case you have been very lucky. Use straps to fasten the hives. One will often suffice but it carries a risk and you want to keep risks to a minimum so use two straps. Put them on several days before the move.

Bill Reynolds of Bilton near Harrogate used to say that queen excluders were honey excluders. In some ways he is right. Bill told me of a system that was an improvement on the metal excluder. I tried it on ten hives down in Strid Woods. In theory it should have worked. The plan was use a piece of plastic such as you would get from a feed sack, cut it to a square leaving an inch clear all round when placed over the brood nest. According to Bill the bees would carry nectar up into the supers more easily and the queen would not pass up the sides to extend her brood nest. Well, it was an exceptional honey flow from the heather than year. The brood nests were crammed with nectar and there were no empty cells for the queen to lay in. All ten queens did what they would do in nature, they moved up into the supers and caused just a bit of a problem but that's beekeeping. I never used his system again but perhaps it should have been half an inch!

I will mention the oilseed rape lest some farmer decides to grow the stuff in this 'rape-free' area. As a honey I quite like the soft-set, smooth textured bland flavour. It is ideal for putting in tea or coffee or even on the cereals but it is better when blended with sycamore or hawthorn.

At not too great a distance there are acres of it. No one can miss those yellow fields smelling of cabbage. It is often present from April onwards and there is no question that bees like it and build up well on it. As most people know the difficulty is the rapid set of the honey in the comb should the temperatures drop. Because there is often a couple of weeks of good weather in the early spring the honey can pour into the hives but like every other crop it is a gamble. The honey is better removed as soon as it appears ripe. Carry this out straight away at the home base or you may regret it.

In some years when the honey has remained on the hives too long and set I have seen piles of supers stacked up containing solid hard combs. The only way in days gone by was to smash up the combs, place the 'mush' in containers in the warming cabinet and strain the stuff off. This is done reluctantly for no one likes to destroy good combs in the spring, they are priceless.

In those agricultural areas where the rape grows there may be a crop of field-beans to follow. They help to keep the bees fed but I confess I never had much of a crop from them. There can be a real dearth of food for the bees in those areas once the rape dies off and the bee colonies can also be reduced in consequence so if you try it, take care.

It is an adventure going there; much like the heather moor trip and with it being at the start of the season the beekeeper is full of hope. I used to imagine how the bees would feel on coming out that first morning after the overnight move and seeing acres of those yellow flowers. As my assistant once said, "I'll bet they think they're in heaven"

I will mention sections but briefly for today most beekeepers go for cut comb i.e. preparing supers of frames with thin, unwired foundation. This makes excellent honeycomb when stocks are large and there is a heavy flow. The heather is ideal for this for it does not run when cut, being a gel or thixotropic. Eight ounce slabs can be cut out by 'eye' to the approximate weight and size or by the use of a template. Try to err on the top-side of weight.

A well-filled one pound section, capped pearly white on each side with the propolis scraped off the visible wood looks magnificent but they are not easily produced in these times of inferior stocks and poor weather. Many are the times I have prepared two or three colonies specifically for sections only to be disappointed at the result. More of the sections have been taken off less than half filled than those capped to the limit and many do not even get started. Without doubt the cut-comb system is better but I am a firm believer in everyone having a go at this other side of honey comb production.

Years ago a honey packing firm was prosecuted for selling English honey which contained Australian honey. Analysis had revealed it contained Eucalyptus pollen and the firm was convicted on this. I mention this because since those days I notice many trees in private gardens of the Eucalyptus variety and wonder how that conviction could stand today?

At the gateway to Farfield Hall, Addingham stands a fine specimen of an Acacia tree. It has been there all my life, since as a boy I collected conkers from the stately horse chestnut trees in the park. The bees work it each spring. There is another by the gate to High House, Addingham and years ago an even larger one cut down by Mrs. Bigland because it became unsafe.

Any over zealous trading standards officer might find a similar prosecution to the one mentioned difficult to prove should local honey be found to contain Acacia pollen from Eastern Europe. Indeed, unless the adulteration was very gross it is doubtful if any such prosecution could be sustained these days such is the spread of foreign trees and shrubs into the stately homes and back gardens of the nation but as I have said, it would depend on the quantity of pollen and other factors!

'After this, when now you see the swarm after emerging from the hives into the open air, swim through the serene summer sky, and marvel at the blackening cloud driven by the wind, mark well:'
Virgil

Chapter 5

Tackling Swarms Various

It is true to say, anyone keeping bees large or small scale will have the pleasure of dealing with swarms from time to time. I say pleasure quite intentionally, for tackling a swarm, right from the first moment that you see it, is a very exciting and challenging experience and something that should be enjoyed. Sadly, with some beekeepers, they see the issue of a swarm as a failure on their part which is ludicrous; others go into wild panic because they don't know what to do, ringing up all and sundry for advice; others size up the situation, work out their plan of action and capture it. As we all know, that is sometimes easier said than done.

Here in the valley the first swarms might appear from about the 12th of May and the last ones towards the middle of July. This does not always follow for I have seen at least two swarms (supercedure) issue in the middle of August while the bees were at the heather and once or twice I have been called to swarms in September, most likely the result of supercedure.

Often towards the end of the season, if a queen is failing, the bees will make queen cells and set about her replacement without swarming, thus holding the strength of the colony together for the coming winter. For a time there may be two queens living in the hive. Sometimes, if the weather is very good as it can be in autumn, when the young queen goes out to be mated a small swarm will issue with her. Left to itself it will often return to the hive but if a novice beekeeper sees it he feels he must interfere. Take my advice and ignore it; give it a chance to return home on its own strength for the survival of that colony may depend on that single queen. With supercedure the workers allow only a few queen cells and they are sometimes larger than those produced at other times of the year, so very different from the large numbers of cells they create for the well-known swarm in the spring and summer.

The beekeeper is rarely aware that supercedure has taken place but if he is one of those who likes to daub his queens with brightly coloured paint and finds that queen is not present in the hive when he examines the colony in the spring, he may scratch his head a few moments before remembering this natural selection of the bees. Of course supercedure can take place at any time. It doesn't have to be in the late season. It all depends on the quality of the queen and the strain of bee. Some do it more than others. I find it a very good trait for those queens so produced are of the best quality. Just think of the work it would save all those beekeepers who are forever requeening hoping to get a better strain if the bees could be guaranteed to do it every season. Perhaps they are evolving in that direction; who knows?

Beekeepers have many and varied ways of trying to prevent swarming, a perfectly natural phenomenon and the bee-books are full of these schemes, some simple, some extremely elaborate. Bee farmers with large numbers of colonies, always under pressure for time in trying to work their bees when the weather conditions are fair, have their own simple plan if they detect a colony is to swarm. They cannot devote much time to each colony. One rough and ready method is to move the stock with the queen cells to the side placing a new box with combs and or foundation on the original site. Into this box goes the queen and some combs with advanced brood but no eggs. The queen excluder is then placed on top and the supers. In effect, it is an artificial swarm with the flying bees returning to work it. The stock placed to the side is given a crown board and roof. The bees will continue to nurse the queen cells in the reduced queenless stock.

On the next visit to the apiary in about a week's time, this box with the queen cells now has a sizeable flying force of its own and it might throw a swarm. To offset this, the beekeeper places it on the other side of the queen-right stock causing its flying bees to join the original colony. It does not always work; nothing is perfect but if conditions are right, with a bit of luck this system works reasonably well.

Happy is the beekeeper that is round and about his apiary on most days, who is able to anticipate when a swarm will issue from a particular stock; who knows his land and where swarms are likely to hang. A quick glance at the hive fronts indicates whether one has come out and a further glance round the usual trees and bushes often shows him where it is for swarms so often return to the same place year after year. Obviously they leave behind some residue or trace that attracts the others and that does make for an easy life for their keeper.

When a swarm issues from the hive it will often circle about for a short spell before settling nearby. Occasionally, if the weather is very hot, it will abscond immediately, or settle high in a tree. Those swarms that go off immediately have already decided on their new home and if that happens to be one of your other apiaries not too great a distance away with a bait hive in position you will often find them hiving themselves. That too is another pleasing aspect of swarming bees, a bit like fishing!

Having dealt with so many swarms over the years one finds the bulk of them fall into various categories and to give those new to the craft a little idea of what to expect, I will list some and explain the general plan of dealing with them. Wear protective clothing at all times. Take no un-necessary risks for the bees may not be your own, may be hungry and bad-tempered.

A Text Book Swarm.

This swarm hangs by a small twig on the tip of a branch in easy reach of the beekeeper. It is a fairly common sight.

Have the skep or a firm cardboard box, a sheet and the secateurs to hand. Place your open container on the ground beneath the swarm on top of the sheet; take hold of the tip of the branch with one hand thus holding the weight of the swarm and very carefully snip it free of the tree with the secateurs. Gently lower it into the skep and wrap in the sheet. Carry it into the shade and leave it until evening when it may be shaken onto a board in front of a new hive of foundation with one comb of food to tempt them to stay and sustain them if the weather is bad. Apart from those hiving themselves, this type of swarm must be the easiest of all.

Swarm on the Ground.

This happens quite a lot, especially with an old queen unable to fly far. Just place the new hive with a comb of food and foundation close up against it on the ground with the entrance nearest the bulk of the bees. They may enter the hive without further ado but to encourage them you could carefully lift a handful to the entrance and give a little smoke. In the evening remove it to its new permanent stand or to a new apiary.

Swarm in a Prickly Bush.

This is the equivalent of the 'awkward squad' and is fairly common. They are more of a challenge than the aforementioned. Firstly, spread the sheet as near to it as you can then try to dislodge it by giving the main branch a bit of a bang or a shake. Have your skep to hand and try to place it over them once they are on the ground. Sometimes it helps to lift a handful of bees and put them in the skep. Of course, if the bush is fairly open and will allow you to get the skep inside it just drop the bees into the skep by banging the branch. You have to judge what is best at the time but don't massacre the bush with the secateurs if it's in somebody else's garden. The worst scenarios are bees in Berberis, gooseberry, holly and rose for if you don't get stung you will most certainly get pricked.

Swarm on a Wall.

Place the sheet on the ground beneath the swarm with the skep propped up with a stone, light the smoker just in case, and with your gloved hand, draw you finger against the stone at the top of the cluster. Some of the bees should peel off and fall onto the sheet and hopefully enter the skep. Keep doing this with your finger always against the stone. Put no pressure on the bees for you will injure them. You need patience for most of this type of work but if you take your time it usually pays off. If the bees enter the skep leave them half an hour or so to allow the flyers to join them then bundle up in the skep and remove elsewhere until evening.

Swarm high on a tree.

If this is in your own apiary you should have a bait hive already in situ. If bees are showing interest in the bait hive all looks well. If not I suggest you salute the swarm, wish it well and go back to your gardening or whatever. You can't win them all. However, if you are a good thrower, you can weight a rope and lob it over the end of the branch, thus lowering the branch to bring the swarm in reach. Have your wellingtons on!

Dealing with one such high in a sycamore last summer I climbed the tree to a great height having placed a sheet on the ground with a new hive on it. The swarm was way out of reach but I was able to climb above it and stand on the branch. Banging on the branch with my feet the swarm took the long drop, landed in clusters on the sheet and the hive and made to enter. I carefully climbed down the tree and went for a walk in the wood. On my return the swarm was back in its original position so I climbed the tree again and repeated the performance. Three times did I do this and in the end, the bees, to show their gratitude flew away. Who could blame them! I would say they had their hearts set on some other beekeeper's hive at not too great a distance. Nothing is wasted and I was glad I could still climb a high tree without injury.

A Swarm in a ventilation hole.

This used to be a common place for a swarm to take up residence, sometimes at near ground level and sometimes high up in the house wall. Once they have gone inside it is better to leave them alone. I would explain to the householder that I would not kill the bees, that in some beekeeping circles it was believed misfortune would befall those who did and that the damage to the property to recover the bees would not warrant the damage and expense of doing so. As these swarms often died out very quickly the householder would be satisfied. We are beekeepers; not bee exterminators; there are enough of those.

After many experiences in dealing with them, those swarms entering houses through holes in walls or roofs should be left alone. It is a good omen for the householder when they settle on him. If there are any left after Varroa I think the more enlightened person would feel honoured that they were providing a home for bees within their own. I would dearly love to have bees enclosed in an observation type hive in my bedroom, coming and going through a tunnel in the frame. Being well off the ground, no one would know they were there. To be awakened by their scent and gentle murmurings would be a good start to the day and who knows; perhaps we would sleep better knowing they were so close!

These examples should give any beginners a little more idea how to react to a swarm and those veterans may manage a smile recalling some of their own escapades for every swarm tells a story.

When a new swarm is hived on foundation I would say it is one of the few occasions when any beekeeper would be justified in helping it by feeding sugar syrup. You see, if the weather turns sour and you do not feed, the swarm will use up the food reserves it has brought to its new home. If the period of nectar shortage persists it will soon dwindle and die. Always, it pays to feed a swarm to help it draw its new combs. It will also stimulate the queen to lay for she will not lay unless there is an ample supply of food. Of course, if there is a prolonged honey-flow taking place there is no reason to feed sugar but the flows in this valley are not reliable so err on the safe side.

Multiple Swarms or casts.

Mating swarms often occur simultaneously and settle on trees and bushes each with their own mated or virgin queens. These small swarms, unless wanted to make increase, should be gathered up and hived together in the evening. Just jerk them onto a board in front of the hive. Once they have all gone inside they will settle down. You may find a few dead queens outside on the board in the morning. This is to be expected.

Swarm on a Wire Mesh Fence.

Hit the fence to knock the swarm to the ground where your skep or new hive is waiting. You may encourage them to enter using smoke.

Inclement Weather Swarms.

If you have a swarm to hive and it is raining heavily it is foolish to cast it in front of the new hive in the evening as described. Instead, place an empty brood box on the floor with the entrance blocked loosely with leaves or grass and have the new hive with foundation nearby. Shutter the bees into the empty brood box making sure they have all come out of the box or skep then place the new hive on top with the lid on top of that. They will remain dry and you can remove the empty box carefully next day.

Swarm Living Wild Several Weeks.

In good summers this happens more commonly. Evergreen bushes, stacks of empty boxes or just hanging under a thick bough, bees will draw wild combs and try to establish their home. Hollow trees where they would normally seek a home are now scarce.

These swarms are best carefully removed to a proper hive if it can be done without injury to you. Just cut the combs at the top and wedge them with hazel sticks in a brood box on a floor. Place the box near the site and remove them to another apiary some miles away in the evening with a box of drawn combs over them to speed up the transfer from the wild combs. If you place them near their old site the flying bees will be lost in their search for their old home.

It may be the start of the next season before you can remove the wild combs so before you start the operation you may consider if it's worth the effort. Most swarms ungathered by the beekeeper die out within weeks and those trying to live in the open are vulnerable to various pests such as woodpeckers.

Swarms in Chimneys.

If the fireplace is still open to the chimney, get a good fire going then place plenty of damp lawn cuttings of leaves on it; something to give lots of smoke. This rarely fails and I have known swarms do a vertical take off and disappear for good.

Regarding any swarm in a dangerous position, always consider your safety. Many of us have recklessly put life and limb at risk when young for that is what young people do and why some young men do not survive. If you are going to risk your life, do it for something more worthwhile than a swarm.

The following little tale is of a lady who used to live at Ilkley and was keen to keep bees. She was very well educated, attractive and just bursting with enthusiasm in trying to grow her own food and so she acquired a colony of bees.

Receiving a telephone call from her in something of a distressed state about a swarm in her garden I decided the only way to understand what she had done wrong was to go up and have a look. On my arrival she was lying in her bikini sunbathing oblivious of the agitated bees surging overhead. Lying on the grass near her was the beginner's book with the page opened at swarming plus an empty tea-cup. She explained in great detail and it crossed my mind how wasteful was an Oxbridge education on some people. I did not mean that unkindly for she was just one of those types one comes across. She was a lovely, well-meaning soul but when it came to practical matters she was lost.

She told me she had followed the instructions in the book. The swarm was hanging in the apple tree comfortably in reach of the ground. The new hive was set up with foundation some twenty yards away. She explained how she had dressed up, gone to the swarm, taken a cup full of bees and thrown them in front of the new hive. This had gone on a long time before she realized the bees she threw down were back in the apple tree before she was. I asked for a sheet and her skep, banged the tree branch to dislodge the good-sized swarm which fell neatly into the skep, then upturned it on the shady side of the tree held up by a stone at the bottom and draped the cooling white sheet over part of it. We then retired for a cup of tea. She was very good at that!

Books are all very well but really as this little story shows, she learned more in five minutes than she had all afternoon from the book. Thereafter she was a confident gatherer of swarms and if she ever knew, I am sure she would not mind me relating this story.

All the above is elementary to the experienced beekeeper but beginners do not always grasp what is required. I hope the former will understand why it is included and that the latter might be spared some grief.

Bill Bielby was the bee advisory officer for the West Riding and specialised in the black bees found in the ruins of Fountains Abbey. These bees lived in the ruined walls and I remember two colonies that seemed to be there all my life. Bill captured a number of swarms and tried to breed from them. They certainly behaved like the British blacks remembered with fondness by some for they were prolific swarmers; conserved their stores in times of dearth and were not overly aggressive. They did not build up to massive size but would gather a modest crop.

Bill received a great deal of publicity about these bees so when a swarm was noticed living in the ruins of Bolton Priory there was great excitement and he was sent for. If only someone had realized I kept an apiary one hundred yards from the ruins it could have saved him a journey. Everyone loses a swarm from time to time and one of mine had chosen to set up home high in the wall of the ruin. He soon realized they were not black like the Fountains Abbey bees.

In the sixties I was for a long time the only beekeeper in Addingham, my nearest neighbours being Mr. Eric Lodge at Burnsall then Frank Beatham with an apiary at Reynard Ings. Frank had his bees on Tom and Kate Mason's land and was a great help to many over the years. The farm cottage was one of the oldest in the valley and had several bee boles in a wall at the rear. Kate would have liked me to have placed one of my skeps there but I never got round to it.

Like Frank, once you become known as a beekeeper, you are sought out to assist the general public with everything from wasp nests, bumble bee nests and genuine swarms. This was all done free of charge but since then the nation's attitude has changed. Very few people do anything for nothing now, even with the teaching of bees. Perhaps they will value the instruction if they pay for it. I do hope so.

Author's 1980's skep apiary

'Let him who makes such things his care, himself bring thyme and pines from the high mountains, to plant them far and wide about their hives:'
Virgil

Chapter 6

Heather-Moor Adventure

For those intent on this gamble make sure you have permission to site your hives and prepare everything well in advance. Remember, you may be working in partial or total darkness. Bees fly late in August and some warm nights will fly to the edge of darkness, especially when dealing with a flow from the balsam. There were times when I felt badly about moving the bees from a secure harvest in the warm valley bottom to an uncertain one 600 to 800 feet higher up on the hills some five miles distant. Some of the heather slopes are at 1200 feet above sea level in various parts of the valley and these can be very inhospitable in bad weather causing great shock to the bees if it coincides with low temperatures.

Flat boards were already in situ (pallets are alright but take up more room). Once again, if it is a warm night I would find thousands of bees clustered up the outside front of the hive. This would delay things no end for it meant smoking them down. If too many were thus clustered I have known me to leave those hives behind and deal with them another time. Working in darkness is a tricky business.

Foam rubbers are the quickest, easiest way to shut bees in but sometimes the foam gets moved in transit and the bees get out by the thousand. Also, use wire-mesh screens instead of the crown boards. If it is but a short distance the green woven plastic mesh available from garden centres cut to square works very well but if you snag it the bees will escape.

Always keep you veil and gloves to hand and wear wellington boots for bees do tend to walk up your trouser legs in the darkness. Sometimes after a busy night with the bees I have wearily climbed into bed and in the middle of the night found the odd bee in with me. A sting in the middle of the back when fast asleep makes one think of a sudden heart attack so be warned, it gives quite a fright.

Once the bees are set down and the roofs replaced give them a few minutes quiet then whip the foam out of the entrance. Do this quickly and get onto the next, dropping each one as you go. There will be a few bees on it and you will only waste time trying to remove them. If dropped in front of the hive they will find their way home. Once all the foams are out use a torch and quickly check and count each one to be sure they are free. Bees can suffocate if left enclosed and that is a disaster. Collect the foams next day.

Midges are often savage on these trips so take measures to protect your self and mind there isn't a bull in the field. I once overlooked that fact and had some fun and games.

When all the chosen stocks are at their new site you could, if you were wealthy and had the time, clear off for a holiday. On the other hand, within a week you are likely to check your hives are still upright with the sheep and unless you are very strong-willed you will not resist the temptation to peep in at least one of the stronger hives to see what the bees are doing in their supers. No one can condemn you for that. That first glimpse of amber heather nectar in freshly drawn-out billowing white combs is always a big treat but don't disturb them too much. Let them get on with their work; the flow doesn't last and they must work while the sun shines. Instead, you could lie down on your back (not too close to them) and watch the aerial activity. It is a bit humbling, especially when you think that by the time the flow is over in about three weeks, all those bees you're watching will be dead, literally worn out by work.

Don't ever believe that nonsense that 'hard work never killed anyone'. It's a real whopper of a lie and I'm not just talking about bees. Perhaps I should give a word of warning about lying on your back near a large number of bee-hives. Whilst the bees will likely show no interest in your presence, remember they will void themselves on the wing and you may get spattered.

Once you are into September and the heather is starting to go back. You'll notice it browning. If the honey is reasonably well capped and the bees have slackened off, remove the honey. I used two ways; one the well-tried clearer boards which, if the nights are very cold will clear the bees from the supers in twenty-four hours. If much of the honey is uncapped and it is still quite warm they may not clear. In that case I would have an empty super and a cover to hand and shake the bees off every comb. It works better if there are two of you for this ñ one to shake and one to place in and cover the box. It's surprising how quickly you can clear an apiary this way. Speed is essential or robbing will start. Place the full box in the vehicle and close down to prevent bees entering then get straight on with the next. It is all a bit brutal and you will get stung so wear all the gear you need to protect yourself.

The advantage of this system over the use of the clearer board is the speed of it and the knowledge that the honey is home and into the safety of your building. With clearer boards you are always at risk of an animal dislodging one and the honey getting robbed out or worse still, if there is a rogue beekeeper about, he may get a good crop of your honey just for the cost of lifting it into his vehicle. Remember, sadly there these people about, bees and honey being at such a premium. Another disappointment that can happen when using clearer boards is that the beekeeper does not notice it does not fit his hive correctly, leaving a gap that bees can enter. He may find the honey has all been removed downstairs in anticipation of his theft.

Once the honey was all in store still in its boxes I personally could not rest until the bees were safely back in the home apiaries before the weather broke. If the rains came it could become a nightmare and in those conditions accidents are more likely to happen, like getting stuck in mud in the middle of the rutted field or going off the track into a ditch. All these things happen. On occasion I have had to use a tractor and trailer to get the bees out. If it has been raining and wet and cold I have risked moving them in daylight but more often it is done at night so always have a good torch handy but don't shine it to the bees or they will rise to it. Keep them in the dark.

Once home, place the roofs back on and remove the foams. Double check the foams are out then go away to a well-earned night's rest.

Once your strength has returned, start extracting the harvest and the best of British.

The above is how it was. It is a very different story now partly because of the physical effort but mainly because of a complete change in attitude to the way I keep my bees.

I would no longer wish to migrate bees anywhere, believing it is bad for their health and heavens above, if we can do anything, no matter how simple, to help them on that score we should do so. Static apiaries seem one way but who will be put off striving for that pot of gold just to improve the health of their bees. Some may, which is why I mention it. The British used to be so kind to their livestock and bees are livestock even if they are and always will be 'wild'!

Here is a word of advice on the matter of communal apiaries. They are not a good idea in any shape or form, whether at the heather moor or in the valley bottom and the reason is this. Every beekeeper works his bees differently and at different times. If a number of beekeepers keep stocks on a site, unless it is well-regulated (and who wants that with its shades of the Soviet?) the bees will be kept on permanent edge by interference. If there are weak colonies present they will receive undue attention from robbers and may be killed off. Remember, stressed bees get disease.

If anyone feeds sugar to any one colony it may send the rest into frenzy.

You will see them testing the security of all the hives. The sound they make when so engaged is not a nice one but is unmistakable. It can really become a nightmare and result in colonies having to be removed at short notice because of the emergency so created.

Even sharing an apiary with a single person can produce just the same problems unless visits are co-ordinated. It is easier to do that with two but with any more it is nigh impossible.

From my observation of such communal apiaries they ultimately lead to the spread of disease and great animosity between the members. That should be avoided for enough are the problems of the day.

Antony Kirby's Model Apiary

'He, therefore, was the first to abound with pregnant bees and numerous swarms, and to strain the frothing honey from the pressed combs:'
Virgil

Chapter 7

Dealing with Honey

The harvesting of a crop of honey is almost as exciting as dealing with a swarm. If it be at the end of the season it is the culmination of your work with the bees and should be a joyous occasion. It most certainly will be a sticky one. Firstly, I will mention how we used to extract honey when we did not have access to the modern contrivances such as presses and the centrifuge.

Imagine a warm kitchen with newspapers on the floor and half a dozen supers of honey stacked near the table; on the table sits a deep bowl or tray and in that a square yard of course, unbleached linen scrim. You take a comb of honey and set to with an ordinary table-spoon in scraping the honey off the comb, one side at a time, into the linen prepared to receive it. When the bowl is full you will gather up the corners of the linen and tie them with a strong cord, creating a loose bag. From the knob of a cupboard, or a strong hook in the beam on the ceiling, you will extend another length of cord. You will carry the receptacle containing the bag of honey across to this cord and raise it up above the bowl to allow it to drip at its own speed. Do make sure all is secure or you will have a mess on your hands (and feet maybe). Seriously this rough, rustic system works fairly well.

If the kitchen can be kept warm all night the easier the honey will flow. You can help it a little by occasionally twisting the bag. It is a fairly slow process but anything worth having doesn't come easy and this is how it was for us. The same procedure will work for the heather honey just as well though you may have to twist the bag a bit more to force it through. Do lick your hands well rather than wasting honey by washing them!

The business of jarring the honey is just as sticky for it means using a jug to ladle it from the bowl to the jar. If done on a tray nothing is lost and you can have the spillage for your porridge.

From this primitive system we advanced a bit by acquiring a small heather press which would take three or four complete combs at once pressed out in the usual linen scrim.

For the spring honey we had an ancient centrifuge which took four shallow frames or two deep frames. It operated with a greasy chain and a winding handle and was back breaking but it was an improvement on the first described method and if you have more than two hives you need something a little better though the cruder method saves a lot of expense and with the right companion is a bit of fun and less like a conveyor belt.

In former times a bucket of water would have been kept nearby to wash hands and everything else in and when the operation was over that water would go towards making that delicious ancient British drink- mead.

The more honey you produce the more you will want modern equipment. There are masses of this and it doesn't come cheap. This is where you should perhaps ask yourself, would it be more sensible to look at other people's bees and buy honey from a reliable source such as another beekeeper or the supermarket.

Some Beekeeping Associations loan out equipment for a small fee but it is not always convenient and with just a couple of boxes of honey you could likely have it extracted by the time you've waited about or gone to collect it. Being independent in beekeeping is a suitable thing to aim for.

For the extraction of heather honey the Mountain Grey press was superb and for those who still use it still is. It is sturdily made and also useful for pressing fruit such as apples and pears. The procedure is similar to the first described method, i.e. scraping the honey from both sides of the comb with a table-spoon or a specially made stainless-steel scraper which fits between the top and bottom of the frame. This is much speedier that the spoon though you will unavoidably break the odd comb or two. The 'mush' of wax and honey should then be poured into the linen scrip set in the well of the press. This is carefully folded over into an envelope; the press top closed down and fastened with the pin. All you have to do then is screw her down but do make sure you have placed a bucket underneath from the outset or your honey will be on the kitchen floor. You may smile but believe me I've seen it done many times.

The alternative to the press which I used very successfully for many years was the spin dryer (no pump). It needs adapting by fitting a flexible perforated stainless-steel drum-like contraption with wooden laths every two inches fastened with brass screws or something that will not rust. The purpose of this is to allow the honey to flow once the spinner is in motion. The linen-scrim bag containing the 'mush' of honey and wax is placed inside the spinner. Make sure the bucket is under the spout before you activate the spinner by lowering the lid.

With heather honey, I found it preferable to fill four 30lb containers by scraping then to place these in the warming cabinet for a few hours before trying to spin the stuff. Do not get it too warm or it will stick to the linen bag and restrict the flow. It really needs just an airing. If it's reasonably warm you can, if you wish, strain it straight into the settling tank. This saves a lot of time. Heather honey from the spinner needs a bit more time to settle for it has taken in a lot of air by spinning so do not rush to bottle it. If you do your jars will look a little like glasses of Guinness.

You see, this system does become a mechanical, conveyor-belt, no time to rest, process. I suppose it depends what you want to do with your life!

The honey is placed in 30lb drums and placed in a warming cabinet for the night. In the morning it is run through a fine nylon mesh over the 'settling tank'. There it is left for twenty-four hours for the scum to rise and once this has been scraped off it can be bottled. The scum can be used on cereals for breakfast for it contains much pollen and other goodness, including small particles of wax which it is said, is good for the heart.

For the half-serious beekeeper, one who is a bit more business-like, a stainless-steel extractor that will take nine shallows and three deeps is most useful for the runny honey. The types of extractors are now legion so do not rush into it; get some advice from someone who knows.

Also at this stage you need a decent-sized stainless-steel settling tank. If you intend to keep bees a long time it is invaluable.

Supers of extracted combs should be stacked on top of each other on a shallow tray to catch dripping honey. In the evening, at the first opportunity, they should be returned to the hives to be cleansed by the bees. Make sure crown boards are correctly fitted to prevent robbing. If this is autumn this is a dangerous time. Do not attempt this in daylight unless it is cold, wet and overcast.

It should be apparent to anyone extracting honey that it has to be done where the bees cannot find you. If there are hives in the area the inmates will come looking, even round the house doors and windows so be warned.

As regards the pressing of heather honey, years ago I knew a good beekeeper called Norman Stewart of Aberfeldy. He worked mainly for heather and he had a large number of hives. This meant he needed a large press and he had just that. It comprised a huge metal, half-rounded tank with the tap at one end and over this, set vertically was the metal plate and above that the screw. Between were beech-wood slabs slatted with half-inch laths. Each took a whole deep comb wrapped in muslin and the whole thing when set up properly would take twelve. Once it was filled, the screw part was lowered to the horizontal over the tank and the serious screwing started. He and his wife could process a lot of honey in a day.

He also produced some good quality heather wax and I remember I was so impressed with it that I purchased some for Peter Hewitt of Haworth to make foundation. At that time it seemed every beekeeper worth his salt was making foundation. I even had foundation making included with one young lad who came to take his Duke of Edinburgh Award test in beekeeping. He made a fine job of it too.

All this is good fun and it shows we can be independent if we have to be but then we go on to learn something else and there is no end to it. Perhaps this is why so many beekeepers live to a ripe old age.

Another fabulous heather honey press belonged to a Mr. Anderson of Forres, on the Moray Firth. In this case he had made it himself using the screw off an old tipper lorry from the air base at Kinloss. Fitted to this was the wheel of a mini motor car. Like the one already described it worked with slabs of wood with laths across to allow the honey to run from the muslin parcels containing the combs. The only problem with these large presses was the storage. They were only used for a short period of time but took up a lot of room for the rest of the year.

Jack Nicholson, a good friend from Denton had another unusual press. This comprised a metal drawer beneath the pressing arrangement which could be pulled out for removing scum. Like the other two it had a tap to flow the honey into tubs. Much smaller than the above two it worked much the same with slabs of slatted wood and cloths to hold the combs. It was a steady, slow job using it and it made one appreciate Mr. Eades Mountain Grey model.

'And mindful of the coming winter they experience toil in summer and lay up their acquisitions into the common stock.'
Virgil

Chapter 8

Natural Comb Beekeeping

Honey bees have been on the planet with a near perfect social order for millions of years before our ancestors crawled out of the swamps. This has been established by expert examination of fossils in stone and amber and is a widely held view. During that time they have managed to survive and for several thousand years they have been brought under the control of man who learned how to keep them to his advantage in the need for honey and wax. Bees were revered for their health-giving food and the wax for a variety of uses, from embalming of the dead, to providing candles and assisting in writing.

In those times the bees were kept on natural wild combs inside wicker, clay-covered baskets, straw skeps or hollowed-out logs. It was like this until late Victorian times when Huber's invention of the bar-framed hive became fashionable. This was quickly followed with the manufacture of embossed foundation. Square-shaped hives became the vogue and still are. Gradually the British skeps passed into history and since then honey bees have not been permitted to live on their naturally drawn combs as they had for millions of years. From my humble experience, the keeping of any creature in un-natural conditions for any length of time leads to trouble. Perhaps time has caught up with the modern beekeeper and this would partly explain why the honey bees are not healthy, happy and wise as once they were.

The un-natural way of living leads to stress and stress leads to disease. I think that is accepted by most people. One has only to glance around the world to see this being born out. The trouble is, even the beekeepers are in such a hurry they have no time to give it much thought therefore they continue with what the modern books and teachers tell them. These systems may work passing well for the bee-keeper but no one is able to ask the bees about it. To test their reaction you have to study them over many years.

Even forty years ago our bees built up much stronger, threw bigger swarms, were more vigorous, managed to live comfortably in the wild in old trees, walls, roofs and chimneys and produced greater crops of honey and I am not wearing rose-tinted specs. Colonies occupying three brood boxes and four or five supers were not uncommon and the flight of bees entering and leaving the hives was awesome.

In those days are main problems disease-wise was Acarine and Nosema but always some of the bees survived those periodic epidemics. The appearance of the Varroa mite really put the cat among the pigeons, killing off completely vast numbers of colonies in different parts of the country. One expects the bees to get on top of it in time but apart from dosing them with doubtful manufactured preparations the average beekeeper seems to be doing very little to improve their plight.

Surely the time is ripe for keeping some colonies on natural combs once again. Who knows, perhaps we will produce stronger, healthier bees. Remembering my own skeppists' days how much they were admired by all who saw them lodged underneath a cosy lean-to roof a yard off the ground. Now I have more time it is my hope to get back to that old-fashioned way, not so much to produce honey but to produce healthy bees.

Well-made bee skeps are available but they are expensive and usually they are only used for catching swarms prior to them being hived in the conventional wooden hive. Of course, you could hive a swarm in a brood box without frames, placing a queen excluder on top beneath a crown board. The chances are the bees will draw satisfactory combs beneath the excluder. You could, if you wished, place conventional supers with frames over the top if you were desperate for honey. Bees in such a box would suffer no disturbance to their brood nest and in effect would be in the same position as skep bees and being on their own combs, they would be living as nature (not the beekeeper) intended them to live. If it hasn't dawned on some of our fellow enthusiasts, nature generally works to a clearly defined plan, even if we do not see it.

Another ideal home for bees is the wicker or willow basket sold in some furniture shops. Some are not too expensive and when coated with a mixture of clay from the ground mixed with animal dung they make strong hives and bees love them.

As with the skeps I place cross-sticks of hazel through the sides to help strengthen the wild combs. When they are newly made they are vulnerable should the hive be knocked over by something but these sticks make the whole thing very secure. If a swarm is hived in one during a honey flow the combs are soon drawn down to the bottom. In a time of dearth, rather than let the bees die, it is easy to create a hole in the flat top and place some kind of feeder over containing sugar syrup. This would be only an emergency measure to get the combs drawn and not standard practice.

Placed on a rough flat board with a groove gouged out at the front to allow the bees to run inside they should be well-able to evict intruders bent on robbing them. There are various ways of providing them with a suitable floor.

It must be stressed here they should not be left out in the open but under some cover. I have successfully placed unused brood boxes over them for the winter and popped a conventional lid on top. This also helps to keep the mice at bay.

You must realize that these alternative methods of keeping bees would not even be considered if I thought we were completely on the right lines with the modern hives and all they entail. Just suppose we have got it wrong. How long from Huber would it take to manifest itself with an insect like the honeybee? Quite a while, I should imagine. Perhaps that first suspected virus known as the 'Isle of Wight Disease' was really the gypsy's warning that all was not well with the new-fangled hives?

I would like to be around long enough to see an improvement in the health of not just the British bees but the world's bees, however it comes about. If a few beekeepers could carry out experiments along with their normal activities it might remind us what we have forgotten or missed completely. Blundering on is not likely to do anything much for a long time if ever and perhaps the bees, like us, do not have that time!

Remember forty years ago there were thousands of wild colonies all over the country living healthily alongside established apiaries. Then came Varroa and most of those wild colonies have gone. A few re-establish themselves for a season or two until the Varroa mites bring them down but they do not seem to last. This seems to tell us that whatever system of beekeeping we practice we will have to continue with some Varroa treatment, whether of the prescribed chemical type or the organic type as I have mentioned. I confess to some satisfaction that I am using the humble wild garlic from the woods as our forebears used it for us for a variety of ailments. It may follow that the mites will build up resistance to garlic as they have with other treatments. Time will tell but until then I will continue to use this safe, easy, free measure of control at the same time as looking for safe alternatives. In the end, given time, the bees will overcome this pest just as they have others and they may get a respite before the next problem comes along. May it be a long respite!

Those who choose to keep bees in a natural way are at danger of being maligned by those entrenched in their ways or those with a vested interest in keeping bees the way we do in the bar-framed, square-boxed hive. It is big business and this last couple of years has become even more so. This increase in the number of colonies just after we have lost so many to Varroa and other ailments may be a mixed blessing. Surely, it would have been wiser to have got on top of those problems first. If they were all in the possession of experienced beekeepers instead of beginners I would feel easier about it but the media has whipped up such an interest and now we are in this unknown situation. Most definitely the bee inspection officers are going to be kept busy getting round them and I wish them well on that. It may be that we will have to revert to former times when very experienced members of associations carried out that work rather than government officials but I doubt if they would want to do it for nothing in these times.

Coming back to colonies living on wild combs, having personally dealt with many living in trees that had to be felled, or stricken in a gale or by lightening; or in cavities in walls or under floorboards, I can tell you this, they all had massive surplus stores to carry them through the winter or even through a whole year should famine occur. Some of that honey was crystallised and very old as were some of the combs but there were always clean new combs as well, replacing others damaged or worn out and attacked by moths and other vermin.

A good store of honey over a colony is vital to its morale and well-being; it is its security; it's insurance against hard times. So many colonies under the care of keepers of bees find themselves living hand-to-mouth at subsistence level that they scarcely can look to the future. For many, where novice beekeepers have taken their stores, there is no future. Every year we hear of young stocks dying out from lack of food. These deaths are avoidable but with some people they are repeated year after year. Already this winter I hear of colonies dying out in the hands of beginners and this is only February.

Imagine a family living in a remote region of the earth have insufficient supplies to take them through the winter safely because some brigands had stolen their supplies. Knowing they were doomed they would hardly be planning for the future.

This is how it must be for the bees in so many cases. Now if the brood area of a colony was equivalent of a box and a half national size, on natural combs that the beekeeper could not plunder, always with plenty of stores and no disturbance, just think of the benefits to the bees.

Occasionally over the years I was brought in to deal with a derelict hive containing bees in some large overgrown wilderness of a garden after the owner had died. There they had lived, unmolested for years, forgotten and abandoned. This was always very interesting to me to see how they had fared without the 'assistance' of man.

I remember one such colony at Nessfield, just across the Wharfe from Addingham. Enquiries revealed it had been left in this state for many years but from the outset I could see it was a very strong colony indeed. It occupied a W.B.C. hive but inside the lifts it was in a national brood box and several W.B.C. inner boxes with no queen excluder present. The top box was filled with sections in heavy wooden frames and these had been bred in by the queen's unrestricted access. On top of that were some squares of old carpet festooned with cocoons from wax-moths and other insects. Mice had not been able to enter.

To the fastidious beekeeper the condition of the stock would seem abhorrent yet here they were, blissfully content, working a flow as they had done year after year, very healthy, of good size and gentle. So much propolis had been used that it was almost impossible to raise the brood frames without breaking them. Every crevice in the hive was plastered with the stuff. I am told that in itself, because of its antibacterial properties it inhibits disease. As with our children perhaps we are being too particular about 'cleanliness' and all the scraping off of propolis and wax is helping diseases to gain hold. It's just a thought.

The colony had lots of honey going back years from the state of granulation and could have withstood a famine of a couple of seasons before it would have gone under. They were not black bees though, looking a bit Italian with their orange-band appearance. It was a shame to interfere with them but the new owners were not keen to have them.

At Manor Park, Burley-in-Wharfedale a large ash tree was struck by lightening and it was found to contain bees. I attended it with Tom Chapman, at that time an Ilkley beekeeper. The tree had been split asunder and a large part of the trunk lay on the ground. It was hollow as old ash-trees often are and it contained a mighty stock of very dark bees. The cavity was huge and filled with combs of brood and honey. Already the poor bees were trying to re-orientate to their changed circumstances but the bulk of them were in deep shock and unsure what to do.

Many cardboard boxes were filled with combs, honey and bees for it was a massive stock due to the room it had to build up and also being in a favourable location for foraging. When I kept placing my arm up into the extremities of the hollow the heat from the bees was amazing and I will never forget that. I felt very sorry for their plight but at least I was able to salvage the bulk of them and place them in a new home.

Another memorable feature of this exercise was the amount of propolis on the sides of the cavity. No wood was visible for the stuff and its scent, coupled with the honey and the bees, would have been in the top range had it been in a perfume bottle.

If you're bees are the healthiest in the land then stay as you are; if they're not then give some thought to this chapter and if you think some of us are over-sentimental about wild creatures and their comfort and welfare that's the way some of us are who have lived and worked with them all our lives; for those who are not I cannot imagine they are even half-alive.

Peafowl enjoying the heat from the bees

'And let the lizards with speckled scaly backs be far from the rich hives, and woodpeckers and other birds;'
Virgil

Chapter 9

Pests and Diseases

I am afraid human beings must top the list but just to show the bees have not been singled out as victims I would remind anyone who needs reminding that here in Wharfedale, these human beings, in my lifetime, have managed to pollute the river Wharfe, render the former rich trout-streams clear of fish and sterile, introduce an American crayfish that is killing our native species, introduced the foul disease of myxomatosis to the rabbit population that kept many of us alive during the Second World War, likely caused the disappearance of the Ring Ouzel and the Cuckoo through the unwise spraying of bracken on the grouse moors in July, a critical time for those birds feeding on caterpillars.

Other harm to honey bees, caused every year, is the inappropriate spraying of dandelions, thistles and ragwort on bright sunny days when there is a breeze. We are told that such sprays may travel up to fifteen miles and are breathed in by every man, woman and child, not to mention the animals. Are all the tumours the human race and animals are suffering from a purely modern phenomenon? I mean, we can't blame Chernobyl for everything.

The number of farmers I hear of dying prematurely from kidney failure and such like; all men deeply involved in sheep farming, sheep dipping, the spraying of the weeds previously described. Pure air is as vital to the bees' well-being as it is to ours yet it is no longer pure. If you visit a mountain region far from the habitation of man you will be reminded what pure air is.

Despite all the improved methods of education there are still some very sad and ignorant people abroad which is why it is better if your apiary is out of view. A friend of mine had his whole apiary burned to ashes by such people. It finished him as a beekeeper. I bought the remnants of his equipment and bees but that was many years ago. Sadly, it cannot be said the world has improved since then.

Other creatures harmful to bees at certain times of the year are field-mice, short-tailed field voles and occasionally shrews. These will try to take up residence in a hive in October which is why you should place mouse guards at the entrances. These three mentioned are endearing little animals but not in the hive. They will cause immense damage and their disturbance can cause the bees to die out. Once the bees have clustered tightly they will even build their nest against them, eat the comb and honey and make an unholy mess.

Only once did I have trouble with a toad and that was when the hive was very near the ground after it had received a swarm. Once on its stand there was no further problem.

The green and greater-spotted woodpeckers are common throughout the valley though the lesser spotted is now quite rare. Never have I had a problem with these birds but I know people who have but they were in Suffolk.

Blue tits and robins take bees but they must be allowed to do so for we all love those birds. Years ago I read in a bee magazine how a beekeeper recommended black cotton stretched in front of the hive to deter small birds. I tried it and caught a blue tit. That was the end of that experiment. Let me see the birds any day. What does it matter if there are a few dozen headless bees on top of a hive now and again? After all it is only nature just like the swifts, swallows and martins taking them on the wing. There is room for all.

The diseases that afflict our bees come and go as they do with any other species. We are currently preoccupied with the Varroa mite and have been for some twenty years. This scourge, so I am told, was introduced into Europe from Asia by a scientist who was unaware of the mites. What a disaster it has been, causing many beekeepers to give up the craft in despair. Various medications have been tried to control it, with some success but alas, where chemicals are used we have no inkling what other harm they are causing the bees.

As regards the Acarine mite which lives in the breathing tubes of the bee, in former times we used a preparation from Nitro Benzene. It was strong-smelling of almonds and we were assured it was quite safe to use. In fact I approached an Industrial chemist who sat on the board to approve such preparations and he doubly assured me it was not carcinogenic. He was wrong and the use of the stuff is now banned.

Remembering the early cases of large numbers of crawling bees with the 'K' wings I was inclined to despatch them at night by quick, hopefully, painless methods. This was my rustic precaution of preventing the spread to otherwise healthy colonies in the same way that farmers cull their cattle to destroy foot and mouth disease. It does seem drastic but disease control sometimes warrants that for everyone's sake.

In the case of Nosema, a disease of the gut of the bee, I felt I never had much success in treating it with the expensive preparations of thirty years ago. The bees either died or they got better. If they died you had nothing to worry about; if they got better you had only two things to worry about, (1) would they remain well. If so you had nothing to worry about; (2) if diseased, you had only two things to worry about and so on!

It is fairly true to say, in Wharfedale and likely in the whole country, Acarine and Nosema are ever present. Given half a chance with a few good seasons in a stress-free environment with a considerate beekeeper, the bees will cope with these diseases in most cases. When it becomes unbearable they will do what we do and die. Remember though, their passing makes room for others just as ours does so don't worry about these things.

Those with a scientific bent are aware of other diseases that afflict honey bees and within the confines of the hive there will be many creatures that live side by side with the hosts. I'll warrant there are many disorders that are not yet recorded. Only when things come to epidemic proportions do we notice something is wrong.

I sometimes see beekeepers messing their bees about with all manner of preparations, treating the colony like it was a sick child. So often they seem to acerbate the problem. Had they left it well alone it probably stood a better chance of recovery for the bees are amazingly resilient.

Chalk brood in Wharfedale is widespread. The damp valley is ideal for the growth of fungus. That's the way it is for this nuisance that lives in the cells of some combs causing the brood to turn to a chalk-like substance, hence the name. In very damp seasons even pollen will become glazed over with a coating of mould. The bees have to live with it and so must we.

As regards American and European foul brood, they are both notifiable and A.F.B. requires the destruction of the colony. The simple match-stick test is usually good enough for the beekeeper of average experience. If a brood cell contains a sticky brown deposit at the bottom of the cell, carefully poke it with your headless match, turn it round until you can see if it is tacky then smell it. A.F.B. is putrid. Carefully dispose of the matchstick by burning and notify the authorities of the suspect hive.

This rotten disease puts any contact hives or equipment under suspicion and it can be very expensive and soul-destroying to see a large bonfire of bees and gear going up in smoke, often because someone has been a little careless in failing to notice the signs.

European foul brood is often treated but many beekeepers question whether colonies are better destroyed. I have no opinion on this. For someone having one or two hives it is of no consequence but for a man with several hundred that would be a terrible thing. Years ago, the treatment of E.F.B. by antibiotics appeared successful but it could mask American foul brood which I discovered to my cost when infected hives were placed near my apiary at Farfield. They had been brought up from Essex, kept at Ilkley for a while then brought to Addingham. It led to one of my biggest disasters in beekeeping. Of course, the chap that brought them would not know of the disease in his hives and this is the worry for all beekeepers. We know the disease can live up to some forty years in old equipment. All it needs is for a nosey bee to pick it up and take it back to its hive for the cycle to be resumed. Believe me; prevention really is better than cure so be vigilant.

Domestic livestock such as cows, horses and sheep can annoy bees and several times I have had hives turned over by sheep. The only solution is to erect a stock fence around them.

I once had a pea-hen nest behind a beehive and when the chicks hatched I found one of them dying from a single bee-sting to its face. It was paralysed and I can only assume it had walked in front of the hive. One of my geese was also very ill for a time from a single sting to its face but it did eventually recover. Bee venom can be very powerful if you receive the full dose and my only advice is not to annoy them too much and so keep the dose low.

Strong winds can often blow the roofs off the hives and on occasion blow hives completely over. Where they are so vulnerable some protection is necessary. In the past I have used hurdles.

For many years I looked at a site in a young plantation near Bolton Bridge and was almost on the point of approaching the estate for permission to place hives there. How glad I was that it got no further than the idea for a great storm came, the beck was the highest for years and the point where the hives would have been was part of the river.

One beekeeper higher up the valley bought some bees from me after he had been absent from the craft for some time. Glad to help him out I sold the bees for very little. He placed his bees on the side of the river Wharfe and they were swept away in a flood. Bees are better away from water for it causes damp mist and reduced temperatures. In nature honey bees do not build their nests near water but in dry woodlands or hollow trees in sunny hedgerows.

In very hot summers, those hives exposed to full sun with metal-topped roofs can get to boiling point. This overheating may cause the bees to vacate the hive and hang up the front and sides. To help with this problem lift the lid, place small pieces of stick beneath the crown-board thus allowing a through circulation of air and put the roof back on diagonally. Remember to replace it before the thunderstorm comes. When I say stick I do not mean something so large that it will permit bees to pass beneath the crown board for that would start robbing.

Another method of cooling the hives is to place the large leaves of butterburr on top of the roofs. We used to call these rat leaves but I can't think why unless it's from the days when water voles were common for the two were often close by. Incidentally, the butterburr flowers produce good nectar and pollen early in the spring at the same time as the willow and bees love the pink flowers.

Having seen friend's apiaries be near wiped out by Varroa in different parts of the country despite all the treatments recommended I have watched with sadness their loss of interest in the craft. It seemed the problems were insurmountable and one can understand them giving up completely. Disease is such a terrible thing and for the bees and us very difficult to fight against. Prevention must ever be the way forward which is another reason for writing this book. Even with the moveable-frame hive there is much more could be done for their welfare as outlined.

Even the person with a couple of colonies can take a lot of the stress of 'management' away from his bees. The colony is not a young tree to be uprooted every week to see how the roots are growing. You can teach the bees absolutely nothing but you can torment them and stress them with your activities. Try not to; they will reward you.

For the successful treatment of colonies for Varroa mites I have used the following. Wild Garlic or Ransoms leaves should be placed on the top of the frames in the spring. I use about a handful and crush them up a bit. In the autumn I dig up the wild garlic bulbs, wash them, and then crush them, placing them on top of the frames about the brood nest. The bulbs are small at that time of year so I use about five or six.

If there is an infestation there will be a large drop of mites in twenty-four hours and fewer each succeeding day. As a control measure it works for me.

I have passed this information to a number of bodies connected with bees but it seems there is an air of timidity or scepticism in the air. Perhaps if I had patented it, had it put in a bottle with a nice label and charged them £10 a time it would have more credence; oh, yea of little faith!

When you think of it, the bees too will be trying to find a remedy for Varroa. They have been rather good down the years at dealing with such handicaps. My own bees have only yards to fly to work the garlic themselves which they do to some extent every year. This natural gathering of pollen and nectar from medicinal herbs may have more effect on pests than we realise. I've never noticed the honey smelling or tasting of garlic for I do not believe they gather it in such quantities, there being better sources available at the same time, but they certainly gather some and if I am having success with this measure it is reasonable to suppose they may be too.

I have no vested interest in advocating natural comb beekeeping or the use of wild garlic as a Varroa preventative; it just seems common sense to me for the health of the bees. My peafowl eat it all year round, even digging up the bulbs in the winter. At times they reek of it. I eat it in moderation in the spring because we were brought up to believe it was good for us and like the leaves from the hawthorn, it satisfies a passing hunger.

Natural Comb drawn in empty super

But as for your hives themselves, whether they be compacted of hollow bark,
or woven with limber osier, let them have their inlets narrow;'
Virgil

Chapter 10

Hives and Equipment

The commonest hive in Wharfedale is the modified national. The original national hive had four smooth sides being double-walled on the sides holding the frame lugs but it was modified to save labour. Why the boffins of the day settled on its dimensions I do not know. People argue about it to this day and at the time it produced angry controversy. I suppose, as with these modern times, someone had a bee in their bonnet about the matter and was sure he was right. To settle on a hive 18 1/8th square still seems a bit barmy to me, especially if you have to make your own but like the bees, we can get used to anything. I suppose if you cast a swarm into a clean oak beer barrel they would live very happily provided the bung hole was clear. It would be well insulated too and what a sweet smell it would have.

The single-brood box modified national hive is approximately the size of the average English skep and this is likely the reason for its birth. Some keep bees on the single brood-box system and it works for them. Ever concerned for the welfare of the bees I found the single box two small. Ideally a shallow box of honey should be over the queen excluder the whole time and the supers above that.

You will read in some books never to leave a queen excluder on for over wintering because the colony may split and the queen left to die. If we were talking about two deep boxes I would agree but with the shallow box I have never had that happen and can truthfully say, the best over wintered colonies have been those on a box and a half or one brood and a shallow of honey. It makes sense if you think about it for it is the bees insurance for hard times. It is also insulation where they need it, right over themselves. There is no better heat retainer in the cluster than solid slabs of honey. In cold weather do you not wear a cap?

Stands for hives are very important whether made of wood or plastic. Pallets are fine and there was a time (after Canada) when I used them quite a lot. They were about five or six inches off the ground when placed on old bricks and it was possible to put several hives on them though I have now discontinued that. The further apart the bees are in the present disease climate, the better. Currently I am using old plastic milk crates. They will bear the sort of weights produced here but look vulnerable when five boxes high. The most practical wooden stands I have seen in recent years were made by that skilful friend Peter Hewitt to whom this book is dedicated.

The stands are four-legged, strongly made, with an angled alighting board and the standard hive floor fits snugly into it making it virtually impossible to dislodge it from the stand by accident. Of course, wood soon deteriorates and I have gone through many stands in my time for no preservative lasts for ever and stands do get neglected.

Single hives kept on a flagstone with two breeze-blocks to stand on are perhaps as good as you will get on static sites. For my own convenience I kept hives in pairs. Some stands were made for just that purpose but I am doubtful if it benefited anyone much but me. It was done for ease of work, for uniting, should one stock become queenless and to give shelter to each other from winds. Bees often drift and enter the wrong hive when they are close together and that invariably means certain death for the intruder. The main argument against hives being close is the spread of disease so really, if you have plenty of room, keep them well apart.

The floors for this hive vary quite a bit so I will not go into that too much. Sufficient to say after much thought I think the hive with the bee-space the full width is the best. Peter Hewitt uses such floors, does not have to place mouse guards on and never gets mice in; the bees seem to get plenty of ventilation. Those floors with the traditional full width 7/8th gap give problems if the entrance block or the mouse guard is dislodged which it often is and then the mice are in. Always go for prevention and use the narrow gap. Some floors have the narrow on one side and the wider gap on the other (winter and summer) so you takes your pick.

Wooden crown boards constructed to take bee escapes for clearing honey are very useful but ideally you will have a set of clearer boards to be used just for that purpose.

Feeders. If you're keeping bees in the interest of the bees you really should not need this piece of equipment unless there is a famine. The Miller feeder named after an American doctor who designed it is an overall feeder resting on the frames. A wire gauze allows the bees to tap the sugar syrup without entering the reservoir where they would risk being drowned. These are expensive to buy for their brief use and time-consuming to make but we all need something handy for emergencies.

White buckets with holes in the lid will last for years but you need a spare brood box in which to contain it. It is a case of filling the bucket, inverting it over a bowl and lowering it onto the crown board where the feed-hole is clear.

For some of us keeping bees just for the pleasure of it, all this feeding of sugar syrup is outdated and detrimental to the bees. It causes undue activity and excitement at the wrong times sending the bees on the rob; it stimulates the queen to lay when she will have been running down, preparing for the dormant season; if in the spring, it will entice bees outside when otherwise they would not think of it and they will succumb to the cold winds and weather.

It is many years since I fed bees and they are much healthier. If this book convinces you to leave enough honey for their all year survival then it has been worth the effort so give it a try for a few seasons. You have nothing to lose and they have everything to gain.

Roofs. Deep, six inch, metal-topped roofs are the best. The extra weight means they rarely blow off and they give good shelter for the boxes beneath. I have no rabbets or bits and pieces inside mine; such can lead to robbing if there is not a close fit. Once mine were mostly painted green and blended in very nicely with the trees behind. In recent years I have placed eighteen inch squares of King span insulation board on top of the crown boards beneath the roof. This stuff is about two inches thick and helps to retain heat, thus saving honey consumption. One doesn't see the melted patch on the metal top in snow when using it.

I have always believed in plenty of ventilation and still do but much of that can be achieved with the wide entrance as well as the feed hole being open in the crown board. Of course, these squares cover all that but the hives seem dry and there appears to be a real bonus to the bees in using them.

The cold, dry winter such as we are going through (2009-10) is a reminder to some of us how it was when, despite the cold weather and the long confinement, bees tended to winter much better. It is the long, wet, damp conditions that harm bees.

Regarding other types of hive, I have experience in using the white painted W.B.C. (William Broughton Carr) hive still occasionally seen as an ornament in neat gardens and on a static site the bees are very happy to live in it. Having outer lifts it keeps the bees very snug and dry in winter.

The Scottish Glen hive is similarly more suited to a static site and has all the attributes of the W.B.C.

The Cottager hive with its gabled roof and double walls looks attractive but they are rarely seen now. It was a cumbersome hive, requiring two people to move it and was essentially for a static position. Bees lived in them very well and if well-constructed the hives would last over a hundred years much like the W.B.C.

The Scottish Smith hive is the smallest hive in use, mostly in Scotland, is very compact and easy to manufacture. Its frames do have short lugs so interchange with other types becomes complicated. If only the national had been designed on those lines!

Some Bee Farmers use the American Dadant hive but for those concerned for their backs it is the biggest one.

The Langstroth is the most popular world-wide and many bee farmers in Britain prefer it. Its brood frames are larger than the national and I Believe its brood nest is equivalent to the Modified national deep and one half. Again, if only those early British beekeepers changing over from the small skeps had got together with their American cousins?

To clear the confusion that might arise and to sum up on hive types for use in this valley, the modified national with one box as a permanent part containing their honey reserves is without doubt the best, especially when you come to sell it. Strange hives are viewed with suspicion by some and those I inherited eventually ended up on the fire.

Speaking of skeps, it would not be right for me to pass over this ancient type of hive without a few comments yet again. The beekeepers of the Victorian age seemed a might hasty in changing from an inexpensive straw hive to a very expensive wooden one. We know bees had lived quite well in those straw skeps for thousands of years. To us the methods of management may seem a little crude though I confess to seeing many beekeepers with the most modern hives and equipment treating their bees just as crudely.

FRAMES

The best frames for me have been the Hoffman self-spacing both for the brood box and the supers. They can become well-propolised and some beekeepers do not like them because of that. They are also more expensive.

Manley self-spacing super frames provide excellent, wide combs but propolis can become a problem with the full-length contact. Both types of frame being self-spacing do away with the use of metal and plastic ends.

The British standard frames with metal ends are still used by many amateur beekeepers, either because they inherited them or they do not know any better. I have used them widely over the years and had some frustration and nasty cuts from them. The coloured plastic ones do not cut one's fingers but they tend to break when pressure is applies. As I have said already, you can get used to working with anything. The old-time beekeepers I knew were pretty ingenious. I've seen some use nails for spacing but I don't recommend it.

The best hive-tool I ever had was made from an old file by Bobby Walbank, an engineer of Keighley. All he did was turn up the tang and sharpen the edge.

At the other end he squared it off and sharpened it fully across after beating it out a little. It was so handy in getting under the frame lugs with the tang and the sharp end was ideal for scraping things. The manufactured one that I now use is not nearly so useful.

Bobby was pretty ingenious and having watched me scraping heather honey with a spoon he made the perfect tool for the job out of a piece of stainless steel, cut the width of the inside of the frame and mounted on a wooden handle a bit like an old type Dutch draw hoe. It saved some hours of work and is practically indestructible.

I wish smokers were indestructible. The type we all tend to buy don't seem to last any time at all but perhaps I am too rough with them. The metal case comes away from the bellows; holes appear in the lid; the grid in the bottom gets thrown out with the ash and lost. If someone were to make a more durable smoker it might not sell so easily for they would last ten times longer. The best I ever saw was home-made by a beekeeper friend called Dennis Knappy from Silsden. It was a plain device made out of aluminium and was very robust.

Dennis also made me my first circular saw out of our old washing machine motor. It never did wear out though it did tremendous work. It would have been going yet but a friend begged it off me and it nearly took his finger off. It could have been worse.

With any machinery, do be careful; it only takes a moment's distraction to result in an accident whether with spinners, presses or saws.

People are suddenly conscious of their environment and awakening to the damage they have caused it and the other wild-life that shares it with us. That is good for it means if we are given time there is hope. Many wooden hives are made out of cedar wood grown in North America. It takes a long time for such a tree to grow. I used mainly red deal. It was slightly heavier, did not have the rot-resistant qualities of cedar but it lasted a long time. Straw, reed and certain grasses highly suitable for skeps grow annually and no one uses them for anything much. If only people could slow down a bit and have a think we could make these islands a land flowing with milk and honey again as it was when the Romans paid us a visit. Incidentally, I read recently how someone stated honey bees did not come to Britain until the 5th century A.D. We were always brought up to believe our forebears the Celts produced vast amounts of honey long before the Romans came. So impressed were the Romans that they recorded it, still to be found on tablets in the British Museum. Worse lies have been told about the British!

Look at the mono-culture of our meadows where useful flowers are now scarce, meadows once beautiful with flowers, cultivated to this level over a thousand years ago where once the British bees gathered a harvest from flowers gone these last fifty years. From a bee's point of view agricultural land is now more or less sterile and if the hapless creatures venture onto it they are likely to get zapped by spray if it's a dandelion, a thistle or a ragwort flower they choose to land upon.

An hour's drive from the valley is an area where I took some bees to the oil-seed rape. I there saw huge trees felled and hedgerows grubbed out at taxpayer's expense, losing vast acres of foraging area for bees and nesting sites for birds. Within a few years the trees and hedges were being replanted, at taxpayer's expense, of course. How crazy can you get?

One can often read articles in the newspapers and countryside magazines about 'Guardians of the Countryside'. To some people this seems one of the biggest attempts at disinformation since the Nazi propaganda minister Herr Goebbels bombarded these islands with lies during the Second World War. Those living and working in the countryside are not fools; they know what goes on. The tragedy is those who have the means to do the most good, seem less knowledgeable than some of those townie naturalists who cite farmers as being responsible for the decline of so much that were good.

Perhaps economics and survival, when it starts to threaten the land, the streams and rivers on which all life depends should be viewed from a different angle. As a boy, the farmers I knew were happy, contented people though quite poor by today's standards. Their meadows were a joy to behold and never did hay smell sweeter. Bees of all species were plentiful; the streams were full of red and black speckled brown trout and the Wharfe was much cleaner than now. Otters lived at Farfield, horse and field-mushrooms were common-place in so many fields and hares were numerous. I detect a genuine concern among some landowners to these losses so perhaps things will improve for the bees and the rest once they get round to it.

It is the loss of agricultural land to bees which makes the trees so important in the valley. Without them it really would be a desert, even with the heather for that poses certain problems to bees and us as I have outlined.

When I see fields full of dandelions, thistles or ragwort I do not rejoice at the thought of so much pasturage for bees; instead I fear for them at the hands of the man on the tractor with the spray. The bees being so small he probably little knows or cares of their presence as he drives about in his cab with the transistor radio blotting out all natural sounds. I suppose this is why more people are trying to keep bees in suburbia or the cities where there may be less danger to them. At least the trees in the municipal park will not be sprayed even if they do have other forms of 'fall-out' from motor vehicles for instance.

In the village of Addingham there are thirteen mature lime trees; in Beamsley two and Bolton Abbey four; Ilkley and Burley have enough for them to be useful to the bees in July. It has been said a single mature lime will provide food for three colonies of bees. Percy Ogden, the well-known beekeeper helped me plant some lime saplings twenty-five years ago. They are now as tall as a house and flower every year. We don't have to plant the big trees; fruit trees will help and every garden should have them. There are so many ways of making Wharfedale or any other area more bee-friendly and trees are one of them.

On my land there are several horse chestnut trees grown from nuts I picked up on St. Stephen's Green, Dublin, a place where the Irish martyrs fought the British in 1916. They grow very well in this valley, producing the well-known red pollen but also fine sweet nectar that we used to tap from the white flowers as children. I think the red flowering variety is not much good. Horse chestnut is an outstanding tree in appearance but requires a great deal of room. Ideally it should be grown in park land to allow its spreading branches to reach their maximum. Because it blooms at the same time as the sycamore the bees tend to go for the latter which can be a very heavy yielder of nectar.

Another valuable plant for the gardener and the bees is the raspberry. Again, every garden could grow a few canes and think of the gorgeous fruit. Every little helps. The honey from the rasp does tend to granulate quite quickly in the comb.

If bees are as important as we think they are, just think how the ordinary, non-beekeeping public could help by planting a few of the things mentioned if they really believed as we do and cared to do something about it. Canaan need not be the only land flowing with milk and honey.

It seems only right to mention back garden beekeeping for it is yet in the media and people are being encouraged to partake in it. My father kept his three hives at the bottom of his long garden to his semi-detached house. The area in front of the hives was wide open countryside but to his left and right were his neighbours' gardens. Apart from the odd swarm entering his neighbours' property there was never any stinging and never once was a complaint made but they were good neighbours and understood the importance of bees; they also liked honey, the best kind of sweetener.

I also used to keep bees in the garden of my detached house; sometimes as many as eight small colonies along the boundary hedge side. Over the hedge was the public road. In addition there were frequent visitors to the house and children were constantly playing in the garden yet hardly anyone was stung and that was through collisions. It was almost as though the bees knew they had nothing to fear from children.

In the case of my father's apiary, his secret for his bees docility was he never disturbed them after placing boxes on in the early spring and taking them off in the late autumn after the bees had more-or-less clustered. They did not know to fear him. He sat by them regularly and they were used to his presence. One could call him a laissez faire beekeeper and frankly, it's not a bad thing to be.

The bees in my garden were there for pollination of fruit trees and removed before they became too strong. I would not have felt happy at having very strong stocks where the public had access and any disturbance to them caused by manipulation would not have been wise.

Many friends in remote cottages in the countryside kept their bees in the garden but if you are surrounded by fields and woods there is no problem from other people and that is perhaps the ideal situation.

Vicars, priests and schoolmasters were often in that position with a detached house and large garden as was the village constable. It was such people who kept beekeeping alive during the doldrum years between the wars when they were stalwarts of the county associations.

Whether a back garden is suitable for keeping bees will depend on many factors. A garden the size of a postage stamp in suburbia would clearly not be suitable but in some towns and cities the gardens can be enormous and they might be alright. In every case one has to consider the neighbours. If it seems there is likely to be conflict, don't even think about it. Among some simple people a solicitor was someone who loitered about the dodgy streets of a city for immoral purposes but today, on your television screens you will see them advertising to sue people on your behalf should you have a grievance.

Britain is vastly changed from fifty years ago and people are quick to go to law about the silliest things. It is something best avoided.

Whether bees would thrive in a given environment can only properly be found out by trial and error over a number of years.

I have known many people keep bees successfully in their garden but they were experienced men. I would not like to think a novice was keeping bees in a garden alongside me for I have seen terrible conflicts arise from such happenings with others and conflict is something we must all try to prevent.

To have a single colony of bees in a large garden would be little different from having a large wasp nest living there with similar risks from stings. If left alone, wasps, generally will not attack people but if you poke a stick down the hole leading to their nest then you must expect people to be strung. Similarly with a colony of bees, if you persistently examine them by pulling the combs apart, they will not like it and will try to defend the colony by stinging anyone within range.

Another important point to mention about bees being kept in back gardens is so often overlooked. When you get council officials examining the adjacent properties for stains on windows, motor cars and freshly hung washing it is perhaps time to keep a low profile until you find a new home for your garden bees. Bees void themselves on the wing, often soon after take-off from the hive. This brown, smelly substance takes a bit of removing and clean washing so spotted can send proud housewives into a frenzy. I have known officials mistake the bee excrement for industrial fall-out until they were enlightened by more intelligent members of their staff. I can imagine this little problem will become a big one should 'back garden' beekeeping really take hold so let's hope those so inclined will properly think it through. Irate washer women should never be confronted after aerial bombardment by your bees. You have been warned!

Village hall stand in the 80's with assistant Joel

"Hag Head Laith" near Highfield, Addingham looking towards Beamsley Beacon

'They are wrathful above measure and when provoked breathe venom into their stings and leave their hidden darts fixed in the veins and lay down their lives in the wound.'
Virgil

Chapter 11

Beekeeping for health

Entering beekeeping on the scale that I did was motivated by being invalided out of my employment through depression. This was over thirty years ago and I mention it for chance that others, in a similar position with a health problem, may derive some hope and comfort from my experiences, for the bees played a very important part in my recovery. So did a number of human beings, I am pleased to say.

Working closely with nature, in all weathers, in all seasons is the finest way I know for a man to restore his damaged body or mind. It was the only way I knew and instinct saw that I got on with it.

Buying tools and timber I busied myself making more bee hives and all that went with them. In addition I created a massive organic garden on a few acres of land I was able to purchase and planted thousands of trees suitable for bees. A number of friends who shared my enthusiasm assisted in these tasks and I was glad of the company.

Gradually the apiaries expanded and soon I was engaged in hard work from dawn to dusk. In addition I was much sought after to give talks at Agricultural colleges, schools, beekeeping associations and the like from as far away as Suffolk to Hartlepool, Whitby and Lancaster. All this activity connected with bees got me on my feet again and meant I slept soundly at night.

How I loved to be with the bees, taking in their silent wisdom in all they had to offer; joyful at the sight of a swarm or the sound of young queens piping in their cells. To see newly drawn combs, white as ivory filled with new honey from the reliable sycamore or the unreliable hawthorn never ceased to fill me with awe at the magnitude of their labours. How could those tiny insects fill dustbin sized drums with honey? No wonder the old Co-operative Movement chose the bee as their emblem for industry.

All this activity meant I was very fit physically and strong for the tasks of lifting boxes and moving hives. It was always enjoyable and the problems that regularly occurred were solved by careful thought and helped to keep the brain in harmony with the rest of me.

I visited Canada, assisting my friend Colin Pullein on his bee farm. I learned much about managing a large-scale enterprise, working for orchard pollination as well as honey. The climate in British Colombia was dry and warm, making the bees easy to handle. The people there were very friendly and seemed to have originated from every country in the world but especially Scotland. The trip boosted my morale even more and I returned to my own apiaries enthused with ideas, especially with queen rearing.

I brought back several hundred new queen-cell cups as used out there and set to making up the frames to hold them. Brood boxes were divided in three with no communication between and similarly separated on the tops. One of my best colonies was chosen to provide the eggs and others were made extra strong to provide the nuclei by use of the Demaree system. When the cells were sealed the nuclei were made up good and strong with plenty of food and bees. It is important not to have any eggs or young brood in those boxes for obvious reasons. Those bees must not be able to start a queen of their own or they may kill off the one you have introduced.

The ripe cell should be inserted between the combs from above and the whole thing should be closed up. It is better to have a special apiary for these nuclei to become mated to avoid robbing but remember you will need at least one good colony present to supply the drones.

Once the young queens are mated and there is a normal patch of brood, including sealed brood, they are ready for uniting to other colonies that desire them. For heather production it was always considered good practice to go to the moors with young queens. They are more vigorous and continue laying until later in the season, ensuring there is a plentiful supply of young bees to go into winter.

Having done much in the line of queen rearing I am still of the opinion that the best queens are produced naturally by the bees themselves as when they prepare to swarm. It is common sense really but there are people who would disagree and who go to great lengths to artificially breed the perfect bee.

As I write my Canadian friend tells me of an apiarist losing 70 colonies and dissatisfaction in imported queens is causing more beekeepers to raise their own. Apparently some of the queens from the imported strain don't last the year.

Sitting about in the apiary is not time wasted. Rarely will you be unaware of the delicious aromas from the hives and that alone is health-giving. Coupled with the scent of the trees, the flowers and the earth itself, especially after a shower of rain, the apiary is a good place to be for anyone who has suffered illness on any scale.

Sometimes I think how marvellous it would be if our rich country could provide special health farms for those who need them where beekeeping, horticulture and the study of wildlife would play a part. To be in that environment for just a short time must be beneficial in helping to restore people back to a full working life again.

I was incredibly lucky when my misfortune struck for I already had many hives and knew exactly how I could help myself. Not everyone is so fortunate. Perhaps in time we humans will harness our resources like our wonderful bees and bring about the changes we only dream of. Miracles do happen all the time in beekeeping for those who are observant such as queenless colonies suddenly becoming queen-right; colonies on the point of starvation suddenly finding a surprising nectar supply and prime swarms in high trees deciding to return to their hive for a few more days or with a bit of luck, flying away!

If there is any truth in the current global warming theory the prospects for beekeeping in Wharfedale look so much brighter. The Yorkshire dales are renowned for their dampness through rain, mist and overcast skies as the Pennines catch and hold the bad weather from the Atlantic. Wharfedale has long been called 'Rheumatic Valley' by those who have lived here for generations enduring that complain with stoicism as they tend their flocks on the fells. This poor weather no doubt accounts for the low number of beekeepers in the valley during my lifetime and for generations before. It is no accident that the bulk of Britain's beekeepers are in the southern counties where the weather is so much kinder. There, the crops of honey are more certain, the bees build up rapidly with their spring two to three weeks ahead of ours and because of much more sunshine take advantage of those crops.

In these hills and dales the bees snatch at the available nectar between periods of bad weather. For them and us it is very frustrating to see masses of blossom hanging on the bough, dripping with moisture, with low temperatures sometimes lasting the whole period of blooming then just as it dies completely, the weather improves briefly. How often do we see that with our main sources of sycamore and hawthorn or even the heather?

When travelling from Wharfedale towards York one often leaves the valley in heavy rain only to find towards Harrogate that the sun is shining. Looking back towards our hills one sees the dark clouds lying there all day long yet over Boroughbridge the weather is perfect for the bees to forage. On returning to Wharfedale we find it has rained without interruption all day.

Where the weather is good, there you will find the most beekeepers. As a boy it was from Kent to Cornwall and up into the Cotswolds, areas where the bulk of England's fruit was grown. The north of England and Scotland were, on the whole, difficult places for keeping bees successfully unless of course their rations were subsidised by gallons of that white refined substance that we are not advised to eat too much of ourselves.

The hardy dales farmers and their equally hardy livestock have been moulded by the weather conditions for centuries. The black bees that once occupied the dales were similarly acclimatised before the 'Isle of Wight Disease' finished them. So far we have not been able to replace them with anything so hardy. To do so will take a great deal of time on their part by breeding in the qualities that give a stronger bee and a better survival rate. Men will try to hasten this but with evolution there are some things that just cannot be rushed.

To venture on a second book of bees in a matter of months might be considered over ambitious by some. No matter, the book 'Buckets of Honey from Boxes of Bees' is a very different type of book. This little volume is an attempt to pass on to the reader that which I have learned about beekeeping in Wharfedale during nearly fifty years. The knowledge so gained must be worth something to those beekeepers who will come after for you can't buy time. If, from glancing through these pages you glean just one or two tips then I will be well satisfied. I mean, why should you have to stumble and stray in your quest for the perfect apiary, the perfect bee or the perfect suit of clothing that will give you 100% protection? Of the latter I know nothing but that may be due to my careless-ness in the way I use what I have; holy veil, holy gloves and holy suit.

It is well-known that a man is not a prophet in his own country and I would hasten to disclaim any presumption to being such but even I do not like waste and not to write down the little knowledge of beekeeping I have gained would seem a sin. How often do we wish we had made a record of the stories of our elders which would have assisted us in so many fields?

We have limited knowledge of the bees and I would be the first to admit that; we have but a fraction of it but enough perhaps to question whether the total change over to the bar-framed hive with its manufactured 100% foundation and all the clutter that goes with the modern hive has been in the interest of the bees. For the beekeeper it was revolutionary in allowing him to become a true control freak over an insect that had managed quite well without his interference but this is the modus operandi of many human beings when they become too numerous and get the bit between their teeth or the proverbial bee in their bonnet.

Sometimes it is good to try old ways again such as keeping bees on natural combs and I hope some of you will for you may be instrumental in restoring the health of our bees to the way I remember it. Do not be dissuaded by the comments of philistines with limited knowledge who may try to frighten you off this course by saying such colonies cannot be inspected for disease. They can, but perhaps not in the orthodox manner that beekeepers have become accustomed to. Perhaps we should remember there are many human beings and other creatures walking about with contagious diseases and no one has managed to spot them either.

No one wants to see the dreaded brood diseases get out of control even if the bees have lived with them for a considerable time. Perhaps even they would be reduced if the bees were kept on a healthier footing and better equipped to fight back by living wholly on honey.

As with any interesting subject you will always find someone who thinks he knows it all; don't you believe it!

The knowledge of beekeeping was not gained without some hardship and heartbreak but through no fault of my own I was given time and the unique position of being in this valley and taking advantage of the situation with the knowledge I already had. Many people helped me along the way and I am grateful to them. They include the Bolton Abbey estate staff, the landowners, gamekeepers, farmers and foresters, beekeepers and bee farmers, friends, naturalists and strangers. They all played a part too.

John Cape hiving a swarm

'If, however, you fear a hard winter, you both be sparing for the future and have pity on their drooping spirits and shattered state;'

Virgil

Conclusion

For honey production there are better places than Wharfedale to keep bees but for someone happy to keep them for their own sake there are fewer places quite so beautiful. The beekeeper prepared to leave the honey for his bees and take just a token for himself without resorting to feeding sugar should enjoy his hobby without having to worry too much about them and he can be assured that those few combs he has taken will be of the highest quality.

The prospects for commercial beekeeping in the valley are poor mainly because of the weather and the dependence in the spring of tree honey mostly used up by the bees in brood production. The summer months can be very lean indeed.

In August and early September, during the period of the heather blooming, if colonies are strong, the bees can sometimes gather a bumper crop of heather honey. However, this means, in most cases, migratory beekeeping, something I am now totally opposed to for the reasons described earlier of shock in transporting inducing stress and disease. Bees deserve better treatment!

Even at Otley bees will produce a heather blend of honey gathered from the Chevin and upstream for as far as Burnsall and even at Arncliffe, bees are able to gather some heather honey from static sites in the valley. Perhaps beekeepers should be satisfied with the modest crop they will produce for their own needs by this well-tried method and leave commercial production to the south of the country or those areas of the planet that can provide nectar in quantity without harming the bees.

It does seem that the production of honey is quite academic until we have re-established healthy, strong colonies for you cannot obtain honey without them. To this end try and breed or retain stocks of bees that appear to withstand the present difficulties of Varroa and other diseases. Whilst it is a bitter pill to swallow it does seem that their lives are literally in our hands. As it seems human beings caused the problem of Varroa in the European bee population perhaps it is their duty to put matters right. Hasten the day!

THE END

Rough cabin in the apiary wherein live bats, bumble-bees and birds

Lightning Source UK Ltd.
Milton Keynes UK
04 October 2010

160758UK00001B/19/P

9 781904 846567